THE CARTOON GUIDE TO

ALGEBRA

ALSO BY LARRY GONICK

"MY GLANCE THROUGH GONICK'S *CARTOON GUIDE TO STATISTICS* BEGAN WITH PROFESSIONAL SKEPTICISM, AND ENDED UP WITH ITS ADOPTION AS THE ONLY REFERENCE TEXT FOR MY GENERAL EDUCATION COURSE 'REAL-LIFE STATISTICS: YOUR CHANCE FOR HAPPINESS (OR MISERY).'"
—XIAO-LI MENG, CHAIRMAN, STATISTICS DEPARTMENT, HARVARD UNIVERSITY

"SO CONSISTENTLY WITTY AND CLEVER THAT THE READER IS BARELY AWARE OF BEING GIVEN A THOROUGH GROUNDING IN THE SUBJECT."
—*OMNI* MAGAZINE

"GONICK IS IN A CLASS BY HIMSELF."
—JEFFREY WASSERSTROM, *NEWSWEEK* MAGAZINE

"[*THE CARTOON HISTORY OF THE UNIVERSE, BOOK 3, IS*] A MASTERPIECE!"
—STEVE MARTIN

"LARRY GONICK SHOULD GET AN OSCAR FOR HUMOR AND A PULITZER FOR HISTORY."
—RICHARD SAUL WURMAN, CREATOR OF THE TED CONFERENCES

GONICK'S CARTOON HISTORIES AND CARTOON GUIDES HAVE BEEN REQUIRED READING IN COURSES AT BISMARCK HIGH SCHOOL, BISMARCK, NORTH DAKOTA; BLOOMSBURG UNIVERSITY; BOSTON COLLEGE; BUCKINGHAM BROWN & NICHOLS SCHOOL, CAMBRIDGE, MASSACHUSETTS; CALIFORNIA INSTITUTE OF THE ARTS; CALIFORNIA STATE UNIVERSITY AT CHICO; CARNEGIE-MELLON UNIVERSITY; COLUMBIA UNIVERSITY; CORNELL UNIVERSITY; DARTMOUTH COLLEGE; DUKE UNIVERSITY; GIRVAN ACADEMY, SCOTLAND; HARVARD UNIVERSITY; HUMBOLDT STATE UNIVERSITY; HUNTINGDON COLLEGE; ILLINOIS STATE UNIVERSITY; JOHN JAY COLLEGE; THE JOHNS HOPKINS UNIVERSITY; KENT SCHOOL DISTRICT, KENT, WASHINGTON; KENYON COLLEGE; LANCASTER UNIVERSITY, ENGLAND; LICK-WILMERDING HIGH SCHOOL, SAN FRANCISCO, CALIFORNIA; LIVERPOOL UNIVERSITY, ENGLAND; LOGAN HIGH SCHOOL, LOGAN, UTAH; LONDON SCHOOL OF ECONOMICS; LOUISIANA STATE UNIVERSITY; LOWELL HIGH SCHOOL, SAN FRANCISCO, CALIFORNIA; THE MARIN ACADEMY; MARQUETTE HIGH SCHOOL, CHESTERFIELD, MISSOURI; MIT; NEW YORK UNIVERSITY; NORTH CAROLINA STATE UNIVERSITY; NORTHWESTERN UNIVERSITY; NUEVA SCHOOL, HILLSBOROUGH, CALIFORNIA; OHIO STATE UNIVERSITY; PENNSYLVANIA STATE UNIVERSITY; PHILIPPINE HIGH SCHOOL, DILMAN, PHILIPPINES; REDBUD ACADEMY, AMARILLO, TEXAS; ROCHESTER INSTITUTE OF TECHNOLOGY; RUTGERS UNIVERSITY; SAINT IGNATIUS HIGH SCHOOL, SAN FRANCISCO, CALIFORNIA; SAN DIEGO STATE UNIVERSITY; SAN DIEGO SUPERCOMPUTER CENTER; SOUTHEAST MISSOURI STATE UNIVERSITY; SOUTHWOOD HIGH SCHOOL, SHREVEPORT, LOUISIANA; STANFORD UNIVERSITY; SWARTHMORE COLLEGE; TEMPLE UNIVERSITY; UNIVERSITEIT UTRECHT, NETHERLANDS; UNIVERSITY OF ALABAMA; THE UNIVERSITY OF CALIFORNIA AT BERKELEY, LOS ANGELES, SANTA BARBARA, SANTA CRUZ, AND SAN DIEGO; THE UNIVERSITY OF CHICAGO; THE UNIVERSITY OF EDINBURGH, SCOTLAND; THE UNIVERSITY OF FLORIDA; THE UNIVERSITY OF IDAHO; THE UNIVERSITY OF ILLINOIS; THE UNIVERSITY OF LEICESTER, ENGLAND; THE UNIVERSITY OF MARYLAND; THE UNIVERSITY OF MIAMI, FLORIDA; THE UNIVERSITIES OF MICHIGAN, MISSOURI, NEBRASKA, NEW BRUNSWICK, SCRANTON, SOUTH FLORIDA, TEXAS, TORONTO, WASHINGTON, AND WISCONSIN; YALE UNIVERSITY; AND MANY MORE INSTITUTIONS OF HIGHER AND LOWER EDUCATION!

THE CARTOON GUIDE TO
ALGEBRA

LARRY GONICK

𝓌𝓂

WILLIAM MORROW
An Imprint of HarperCollins*Publishers*

HarperCollins books may be purchased for educational, business, or sales promotional use. For information please e-mail the Special Markets Department at SPsales@harpercollins.com.

FIRST EDITION

Library of Congress Cataloging-in-Publication Data has been applied for.

ISBN 978-0-06-220269-7

15 16 17 18 19   OV/RRD   10 9 8 7 6 5 4 3 2 1

CONTENTS

THE CHALLENGE IN ALGEBRA IS TO KEEP IT FUN WHILE KEEPING IT REAL—
THE CHALLENGE BEING THAT REALITY ISN'T ALWAYS FUN. THE AUTHOR IS
INDEBTED TO ANDREW GRIMSTAD, DAVID MUMFORD, HEATHER DALLAS, AND
MARC OWEN ROTH FOR THEIR HELPFUL COMMENTS AND CONVERSATIONS
IN THIS REGARD. SPECIAL THANKS TO MARC FOR SUGGESTING THE
"BABYLONIAN GRAPHIC" TREATMENT OF COMPLETING THE SQUARE.

The Multiplication Table

1	2	3	4	5	6	7	8	9	10	11	12
2	4	6	8	10	12	14	16	18	20	22	24
3	6	9	12	15	18	21	24	27	30	33	36
4	8	12	16	20	24	28	32	36	40	44	48
5	10	15	20	25	30	35	40	45	50	55	60
6	12	18	24	30	36	42	48	54	60	66	72
7	14	21	28	35	42	49	56	63	70	77	84
8	16	24	32	40	48	56	64	72	80	88	96
9	18	27	36	45	54	63	72	81	90	99	108
10	20	30	40	50	60	70	80	90	100	110	120
11	22	33	44	55	66	77	88	99	110	121	132
12	24	36	48	60	72	84	96	108	120	132	144

Chapter 0
What Is Algebra About?

BEFORE ALGEBRA, WE LEARN HOW
TO COMBINE NUMBERS BY ADDING,
SUBTRACTING, MULTIPLYING, AND
DIVIDING ACCORDING TO THE RULES
OF ARITHMETIC. TO GO ON IN THIS
BOOK, YOU MUST KNOW ARITHMETIC!

PIECE OF
CAKE...

IF ARITHMETIC IS ABOUT COMBINING NUMBERS, THEN WHAT IS ALGEBRA ABOUT? TO ANSWER THIS QUESTION, BEGIN WITH SOME ORDINARY ARITHMETIC PROBLEMS...

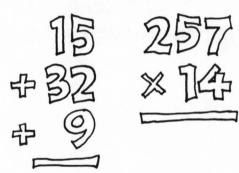
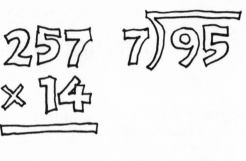

$$15 + 32 + 9$$

$$257 \times 14$$

$$7\overline{)95}$$

AND REWRITE THOSE PROBLEMS HORIZONTALLY, ALONG A LINE:

$$15 + 32 + 9 = \text{WHAT?}$$

$$257 \times 14 = \text{WHAT?}$$

$$95 \div 7 = \text{WHAT?}$$

IN THIS FORM, AN ARITHMETIC PROBLEM IS AN **EQUATION,** A STATEMENT THAT ONE QUANTITY **EQUALS** ANOTHER, BUT WITH A TWIST: ONE SIDE OF THE EQUATION, THE **ANSWER,** IS **UNKNOWN,** AT LEAST UNTIL WE WORK OUT THE CALCULATION.

$$2 + 2 = 3 + 1$$

EQUATION, BOTH SIDES KNOWN

$$\frac{3 + 75}{13} = \text{WHAT?}$$

ARITHMETIC PROBLEM: AN EQUATION WITH ONE SIDE UNKNOWN

ALGEBRA ALSO INVOLVES EQUATIONS, BUT WITH THIS ONE LITTLE DIFFERENCE: THE UNKNOWN ANSWER—THE "WHAT?"—CAN BE **ANYWHERE.** RATHER THAN SITTING ALONE ON ONE SIDE, THE UNKNOWN CAN BE STUCK IN THE MIDST OF THE EQUATION, OFTEN IN MORE THAN ONE PLACE. HERE IS AN ALGEBRA PROBLEM:

$$2 \times \text{WHAT?} - 3 = 11$$

THE PROBLEM IN WORDS: IF YOU DOUBLE A NUMBER AND SUBTRACT 3, THE RESULT IS 11. WHAT IS THE NUMBER?

IN ALGEBRA, WE TREAT THAT THING CALLED "WHAT?" AS JUST ANOTHER NUMBER, TO BE HANDLED IN THE SAME WAY AS YOU WOULD TREAT 1 OR 2 OR 6. (BUT INSTEAD OF "WHAT?," WE'LL USUALLY WRITE x OR y OR SOME OTHER LETTER.)

WE'LL SEE HOW TO MAKE AND USE MANY COMBINATIONS OF LETTERS AND NUMBERS, COMBINATIONS KNOWN AS **ALGEBRAIC EXPRESSIONS.** LIKE HUMAN EXPRESSIONS, ALGEBRAIC EXPRESSIONS CAN BE SIMPLE OR EXCEEDINGLY COMPLEX.

A SIMPLE EXPRESSION

A MORE COMPLICATED EXPRESSION

IN ALGEBRA, THE EQUATION COMES FIRST. AN EQUATION SAYS THAT ONE EXPRESSION EQUALS ANOTHER. THEN WE PUSH ITS EXPRESSIONS AROUND...

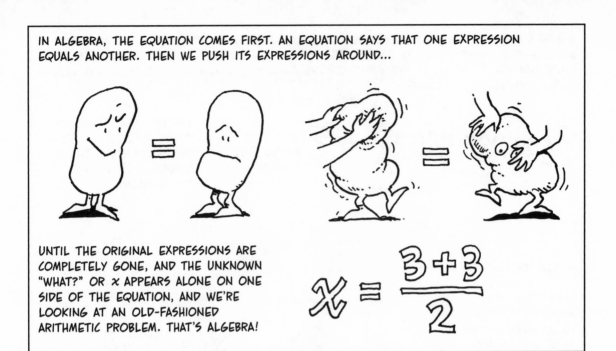

UNTIL THE ORIGINAL EXPRESSIONS ARE COMPLETELY GONE, AND THE UNKNOWN "WHAT?" OR x APPEARS ALONE ON ONE SIDE OF THE EQUATION, AND WE'RE LOOKING AT AN OLD-FASHIONED ARITHMETIC PROBLEM. THAT'S ALGEBRA!

$$x = \frac{3+3}{2}$$

TO DO ALGEBRA, THEN, WE NEED TO LEARN HOW TO "MANIPULATE" OR HANDLE EXPRESSIONS. THERE ARE RULES FOR DOING THIS, JUST AS THERE ARE RULES FOR ARITHMETIC. NOT EVERY MANIPULATION IS ALLOWED!

THERE ARE LAWS ABOUT THIS!

OKAY! FINE!

WE START WITH THE SIMPLEST EXPRESSIONS OF ALL: NUMBERS THEMSELVES. SOME OF THIS MATERIAL MAY BE FAMILIAR, BUT SOME MAY BE NEW...

Chapter 1
The Number Line

NUMBERS HAVE MANY USES, MOST ESPECIALLY **COUNTING** AND **MEASUREMENT.** COUNTING IS THE MOST NATURAL THING IN THE WORLD: THE NUMBERS 1, 2, 3, 4... CAN COUNT ANYTHING, LIKE APPLES, ORANGES, GRAINS OF SAND ON THE BEACH...

GRAINS OF SAND?

I SAID "NATURAL," NOT "EASY."

UM... 2,014,532,578, 2,014,532,579...

THAT'S WHY MATHEMATICIANS REFER TO THE NUMBERS 1, 2, 3, AND SO FORTH AS **NATURAL NUMBERS,** AS IF ANYTHING ELSE WAS, WELL, YOU KNOW, NOT.

EEK!

BUT NATURAL NUMBERS ARE LESS USEFUL WHEN YOU WANT TO **MEASURE** INSTEAD OF COUNT... MEASURE THE LENGTH OF SOMEONE'S FOOT, FOR EXAMPLE.

WHOA! DO **OARS** COME WITH THOSE?

IF YOU SET YOUR FOOT ON A MEASURING STICK MARKED OFF IN SOME UNITS (INCHES, CENTIMETERS, PICAS, ELLS, IT DOESN'T MATTER), THE END OF YOUR TOE MAY NOT LINE UP EXACTLY AT ONE OF THE TICK-MARKS.

YOU HAVE A CHOICE: EITHER SLICE A LITTLE OFF, OR ACCEPT THE IDEA THAT THERE ARE NUMBERS **BETWEEN** THE WHOLE NUMBERS. **FRACTIONS** LIKE 1/2 OR 35/8, FOR INSTANCE, WOULD BE NUMBERS LIKE THIS. OUR IDEA OF NUMBERS HAD BETTER INCLUDE FRACTIONS!

I'M PRETTY SURE AMPUTATION IS MORE PAINFUL THAN FRACTIONS!

WE FIRST LEARN ABOUT FRACTIONS AS "PARTS" OF THINGS. 1/3 OF A PIZZA IS WHAT YOU GET WHEN YOU DIVIDE IT INTO THREE EQUAL PIECES; 2/3 IS TWO OF THOSE PIECES, ETC.

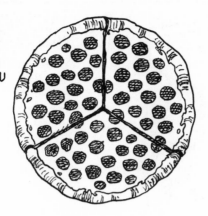

THIS LEAVES OPEN THE QUESTION OF WHAT A FRACTION "IS." IS IT A DIVISION PROBLEM? A NUMBER-SLICE?

AND HOW MANY PIECES OF PEPPER-ONI ARE ON A SLICE???

FOR THE PURPOSE OF MEASUREMENT, A FRACTION IS JUST ANOTHER POINT ON OUR RULER. 1/3, FOR INSTANCE, SITS 1/3 OF THE WAY FROM 0 TO 1. THE FRACTIONS 2/3, 3/3, 4/3, 5/3, AND SO ON, ALSO HAVE DEFINITE POSITIONS ON THE RULER. AND YES, 3/3 = 1, 6/3 = 2, ETC.!

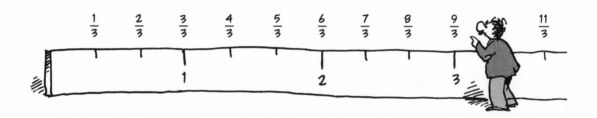

IN OTHER WORDS, **A FRACTION IS JUST ANOTHER KIND OF NUMBER,** A LENGTH, SOMETHING TO MEASURE WITH. EVERY FRACTION, EVERY POSSIBLE COMBINATION OF NUMERATOR AND DENOMINATOR, HAS ITS PLACE SOMEWHERE ON THE MEASURING STICK. IF YOU CAN'T MEASURE YOUR FOOT WITH FRACTIONS, YOU CAN AT LEAST GET AWFULLY CLOSE!

I'M A PERFECT SIZE $\frac{5,651,048}{431,915}$, GIVE OR TAKE...

WHERE DOES IT ALL END?

IT DOESN'T...

WHEN WE GET BEYOND MEASURING BODY PARTS, WE ALSO NEED TO USE

NEGATIVE NUMBERS.

OH, DEAR... I ALWAYS TRY TO STAY POSITIVE!

FOR EXAMPLE...

TEMPERATURE:

EVERY TEMPERATURE COLDER THAN ZERO IS CONSIDERED TO BE NEGATIVE.

I KNEW I LIKED POSITIVE BETTER!

TIME: IF YOU UNWIND A CLOCK DIAL, YOU CAN THINK OF TIME AS MEASURED ALONG A LINE.

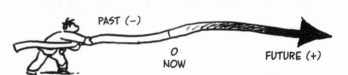
PAST (−)

0
NOW

FUTURE (+)

THE PRESENT MOMENT (OR ANY OTHER TIME, LIKE THE BEGINNING OF A YEAR OR A CALENDAR ERA) CAN BE THOUGHT OF AS ZERO. EARLIER TIMES ARE NEGATIVE, AND LATER TIMES ARE POSITIVE.

I WAS BORN IN −320, AND I'M STILL CONFUSED TO THIS DAY.

MONEY: EVEN **MONEY** CAN BE NEGATIVE! A BOOK-KEEPER TREATS A **DEBT** AS **NEGATIVE DOLLARS**. IF YOU OWE SOMEONE $5, THEN YOU "HAVE" NEGATIVE FIVE DOLLARS, OR $−5.

WELL, AT LEAST I HAVE SOMETHING...

THERE MUST BE A PLACE ON OUR MENTAL MEASURING STICK FOR NEGATIVE NUMBERS. THEIR PLACE IS ON THE OTHER SIDE OF ZERO, COUNTING OFF TO THE LEFT. THE NUMBER 0 SEPARATES NEGATIVES FROM POSITIVES. IMAGINE AN ENDLESS **NUMBER LINE** STRETCHING OFF IN BOTH DIRECTIONS (ENDLESS BECAUSE THERE IS NO LARGEST NUMBER).

NOT EVEN A BAJILLION?

THEN COMES A BAJILLION AND ONE...

THE LINE'S NEGATIVE PART LOOKS EXACTLY LIKE THE POSITIVE PART, BUT GOING THE OTHER WAY. NEGATIVES ARE THE **MIRROR IMAGES** OF POSITIVES.

FRACTIONS INCLUDED!

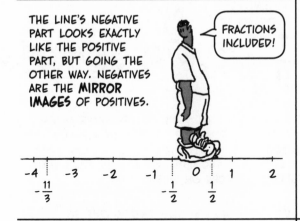

THE **NEGATIVE OF A NUMBER** IS ITS MIRROR IMAGE ON THE OPPOSITE SIDE OF ZERO. IF YOU FLIPPED THE WHOLE LINE AROUND 0, EACH NUMBER WOULD LAND ON ITS NEGATIVE.

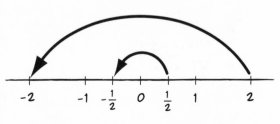

THIS FLIPPING MOVEMENT ALSO SENDS EACH NEGATIVE NUMBER TO THE POSITIVE SIDE. THAT'S WHY WE SAY: **THE NEGATIVE OF A NEGATIVE IS POSITIVE.**

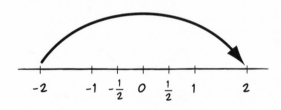

THE NEGATIVE OF −2, FOR INSTANCE, IS 2. WE CAN WRITE THIS FACT AS AN EQUATION:

TWO MINUS SIGNS "CANCEL OUT."

$$-(-2) = 2$$

THE NUMBER LINE HOLDS ALL POSITIVE AND NEGATIVE WHOLE NUMBERS AND FRACTIONS. DO WE NEED ANY OTHER NUMBERS BESIDES THESE FOR MEASUREMENT? IN FACT, WE DO...

AND HERE'S WHY:

AS YOU MAY KNOW, YOU CAN TURN ANY FRACTION INTO A DECIMAL BY LONG DIVISION. HERE ARE 2/3, 5/8, AND 1/7:

```
      .6666...              .625
  3) 2.0000...          8) 5.000
     1 8                   4 8
     ---                   ---
      20                    20
      18                    16
      --                    --
       20                    40
       18                    40
       --                    --
        20                     0
```

AND SO FORTH...

```
     .1428571428...
  7) 1.0000000000
     7
     -
      30
      28
      --
       20
       14
       --
        60
        56
        --
         40
         35
         --
          50
          49
          --
           10
            7
           --
           30       AND SO
                    FORTH...
```

WHEN DIVIDING ONE WHOLE NUMBER BY ANOTHER, THERE ARE ONLY TWO POSSIBILITIES. THE DECIMAL EITHER

TERMINATES (ENDS, STOPS) AS WITH

5/8 = 0.625

OR **REPEATS A PATTERN** ENDLESSLY, LIKE

2/3 = 0.666666666.....

1/7 = 0.142857 142857 142857....

WHY? LOOK AT THE DIVISIONS ON THE LEFT. IF A REMAINDER IS EVER 0, THE DECIMAL TERMINATES. IF NOT, WELL... EACH REMAINDER MUST BE LESS THAN THE DIVISOR, SO THERE CAN BE ONLY SO MANY POSSIBLE REMAINDERS. AS YOU KEEP DIVIDING, YOU MUST HIT ONE OF THEM A SECOND TIME, AT WHICH POINT THE ENTIRE PATTERN MUST REPEAT.

AS FAR AS THE EYE CAN SEE, AND FARTHER!

$$\frac{1}{11} = 0.09090909090\ldots$$

IT SO HAPPENS THAT SOME NUMBERS HAVE AN EXPANSION THAT DOES **NOT** ENDLESSLY REPEAT A PATTERN. ONE EXAMPLE IS $\sqrt{2}$, THE SQUARE ROOT OF 2. (THIS IS THE NUMBER WHOSE PRODUCT WITH ITSELF IS 2. MORE ON THIS LATER!)

$\sqrt{2}$ = 1.41421 35623 73095 04880....

ANOTHER NON-REPEATER IS π, PI, THE DISTANCE AROUND A CIRCLE WITH DIAMETER = 1.

π = 3.14159 26535 89793 23846...

THESE NON-REPEATERS ARE CALLED **IRRATIONAL** NUMBERS, AND THEY HAVE THEIR PLACE ON THE NUMBER LINE, TOO.

GETTING CROWDED DOWN THERE...

BY THE WAY, "IRRATIONAL" DOESN'T MEAN WACKY OR UNPREDICTABLE, ALTHOUGH SOMETIMES IT MUST HAVE SEEMED THAT WAY. AT ONE TIME, SQUARE ROOTS USED TO BE CALLED "SURDS," AS IN **ABSURD**.

YOU'LL GET USED TO ME...

WHAT IRRATIONAL DOES MEAN IS THAT THESE NUMBERS CAN NEVER BE WRITTEN AS A **RATIO** OF TWO WHOLE NUMBERS—IN OTHER WORDS, AS A FRACTION. (A FRACTION'S DECIMAL EXPANSION MUST TERMINATE OR REPEAT.)

1.41421356237309504880168872420969
80785696718753769480731766797379907
3247846210703885038753...
...091...0249...48...05585
...49...935...31...22665
...579995...52782...60571
470109559971605970...3459...2014
728517418640889198...2...230484
3087143214508397...51407
98968725339654633...88296640620615
25835239505474575028775996173.

CLOSE, BUT NOT EXACT!

$\sqrt{2}$

EVERY MEASURING NUMBER, THEN, IS ONE OF THESE:

Integer
A WHOLE NUMBER, POSITIVE OR NEGATIVE

Rational
A NUMBER THAT **CAN** BE WRITTEN AS A FRACTION

Irrational
ANYTHING ELSE

ALTOGETHER, THIS LINE FULL OF NUMBERS IS CALLED THE "REAL" NUMBERS, BUT WHETHER THEY'RE AS REAL AS, SAY, A ROCK OR A PIECE OF CHEESE I LEAVE FOR YOU TO DECIDE...

THEY GIVE ME A HEADACHE, SO THEY MUST BE REAL!

7.43126...

Problems

1. HERE ARE SOME WARM-UP ARITHMETIC PROBLEMS. EXPRESS THE RESULTS OF DIVISION AS DECIMALS. **DON'T USE A CALCULATOR!!** WE WANT TO STRETCH OUR "MATH MUSCLES" HERE!

a. 24
+ 7

b. 58
+35

c. 1.563
+ 0.0002

d. 19
× 3

e. 5.7
× 2

f. 5.7
× .06

g. 1.4142
× 1.4142

h. 2)‾50‾

i. 0.2)‾50‾

j. 21)‾110‾

2. USE LONG DIVISION TO FIND THE DECIMAL EXPANSION OF EACH FRACTION.

a. $\frac{1}{5}$ **e.** $\frac{5}{9}$ **i.** $\frac{47}{100}$

b. $\frac{6}{5}$ **f.** $\frac{4}{11}$ **j.** $\frac{22}{23}$

c. $\frac{47}{12}$ **g.** $\frac{3}{17}$ **k.** $\frac{5}{16}$

d. $\frac{3}{8}$ **h.** $\frac{3}{100}$ **l.** $\frac{4}{25}$

3. SOMETIMES WE INDICATE A REPEATING DECIMAL BY WRITING A BAR OVER THE REPEATING PART. FOR EXAMPLE, INSTEAD OF WRITING

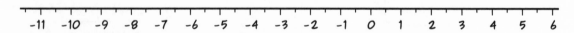

0.010101010101010101010101010101010101010101010101010101010101010101010101010101010......,

WE WRITE $0.\overline{01}$. MUCH SHORTER! USE THIS BAR NOTATION TO WRITE EACH REPEATING DECIMAL FROM PROBLEM 2.

4. CONVERT EACH **IMPROPER** FRACTION TO A **MIXED** NUMBER. (AN IMPROPER FRACTION IS A FRACTION WHOSE NUMERATOR IS BIGGER THAN ITS DENOMINATOR; A MIXED NUMBER IS AN INTEGER PLUS A FRACTION, LIKE $2\frac{2}{3}$. FOR EXAMPLE, $5/4 = 1\frac{1}{4}$.)

a. $\frac{6}{5}$ **c.** $\frac{19}{4}$

b. $\frac{47}{15}$ **d.** $\frac{22}{17}$

5. EXPRESS 3.514 AS A FRACTION.

6. LOCATE THESE NUMBERS ON THE NUMBER LINE: 4.51, $\frac{22}{7}$, $-10\frac{1}{2}$, $\frac{11}{2}$, -3.6

```
 -11  -10  -9  -8  -7  -6  -5  -4  -3  -2  -1   0   1   2   3   4   5   6
```

GIVEN TWO NUMBERS, THE **GREATER** IS THE NUMBER LYING TO THE RIGHT ON THE NUMBER LINE.

THE GREATER
NUMBER
↓

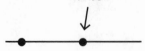

7. WHICH NUMBER IS GREATER?

a. 2 OR 3

b. 2 OR -3

c. -2 OR -3

d. -2 OR 3

e. -350 OR 2

f. $\frac{1}{4}$ OR $\frac{1}{2}$

g. 3.808 OR 3.81

h. $-\frac{22}{7}$ OR -3.25

8. WHAT IS $-(-(-2))$? WHAT IS $-(-(-(-2)))$? WHAT IF THERE ARE 20 MINUS SIGNS IN FRONT OF 2? WHAT IF THERE ARE 35 MINUS SIGNS?

Chapter 2
Addition and Subtraction
(With a Parenthetical Aside)

IN OUR EARLIEST MATH CLASSES, WE LEARN THAT ADDING TWO NUMBERS MEANS COMBINING ALL THEIR "ONES" AND COUNTING, WHILE SUBTRACTING MEANS TAKING SOME AWAY...

WHICH IS FINE FOR NATURAL NUMBERS, BUT MAYBE NOT SO FINE OTHERWISE. TO DO ALGEBRA, YOU'LL HAVE TO GET COMPLETELY COMFORTABLE WITH ADDING AND SUBTRACTING **NEGATIVE NUMBERS**.

BEFORE WE GET INTO IT, THOUGH, WE NEED TO SAY A FEW WORDS ABOUT **PARENTHESES** (CAN'T DO WITHOUT 'EM!).

IN WRITTEN PROSE, PARENTHESES INDICATE AN ASIDE, SOMETHING EXTRA... BUT NOT IN MATHEMATICS!

IN MATH, THEY'RE USED AS **GROUPING SYMBOLS** THAT TELL US TO THINK OF EVERYTHING INSIDE THE PARENTHESES AS A SINGLE UNIT OR QUANTITY OR THING.

DON'T LOOK AT ME LIKE THAT... I DIDN'T PUT 'EM HERE!

SO $2 \times (3 + 4)$ MEANS "2 TIMES THE QUANTITY 3 + 4," OR $2 \times 7 = 14$

PARENTHESES SAVE US FROM WRITING ANYTHING AS STRANGE, CONFUSING, AND STOMACH-TURNING AS

$$5 + -3$$

PLUS-MINUS?

UGH.

GAG—

INSTEAD, WE GROUP THINGS TO CLARIFY MEANING AND IMPROVE DIGESTION!

$$5 + (-3)$$

"5 PLUS NEGATIVE 3"

BETTER.

I GET IT!

MY NAUSEA— IT'S GONE!

GROUPING MEANS THIS: WHEN YOU SEE PARENTHESES, DO THE ARITHMETIC **INSIDE** THE PARENTHESES **BEFORE** DOING ANY WORK OUTSIDE. AS WE'LL SEE, THIS MATTERS!

FIRST ADD 3 + 4, **THEN** MULTIPLY BY 2.

AND ONE OTHER THING: FROM HERE ON, WE WILL RARELY USE THE SYMBOL × TO MEAN "TIMES," AS IN MULTIPLICATION. × LOOKS TOO MUCH LIKE x, ALGEBRA'S FAVORITE LETTER.

GO AWAY NOW.

INSTEAD, WE'LL USUALLY INDICATE MULTIPLICATION WITH A LITTLE DOT, · , OR, WHEN WE'RE IN A REALLY MINIMAL MOOD, JUST BY PUTTING TWO NUMBERS SIDE BY SIDE AND USING PARENTHESES IF THERE'S ANY CHANCE OF CONFUSION, LIKE THIS:

$$(2)(3+4)$$

IS × GONE? ARE YOU READY FOR ME NOW?

ALMOST... BUT FIRST...

FIRST LET'S TALK ABOUT ADDITION...

START WITH A FRESH LOOK AT SOMETHING FAMILIAR: ADDING AND SUBTRACTING **POSITIVE NUMBERS.** WE CAN THINK OF TWO NUMBERS AS LENGTHS (2 AND 3, IN THIS CASE) SITTING OVER THE NUMBER LINE.

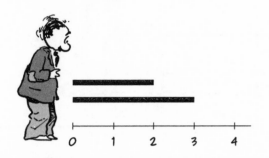

TO ADD THE NUMBERS, LEAVE ONE LENGTH WHERE IT IS (DOESN'T MATTER WHICH ONE) AND MOVE THE OTHER...

TO THE FAR END OF THE FIXED ONE...

AND ATTACH IT, END TO END, EXTENDING OUTWARD. THE **SUM** IS THE TOTAL LENGTH.

NO BIG SURPRISE!

$$3 + 2 = 5$$

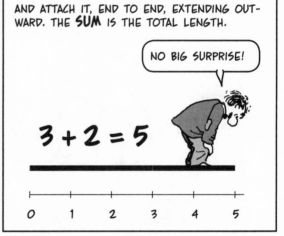

TO SUBTRACT THE SMALLER NUMBER FROM THE LARGER, I AGAIN LAY THE LENGTHS END TO END—BUT NOW WITH THE SMALLER LENGTH **INSIDE** THE LARGER ONE.

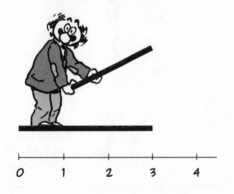

THE **DIFFERENCE** IS THAT PART OF THE LARGER NUMBER THAT DOESN'T OVERLAP. IT'S WHAT'S LEFT WHEN YOU TAKE AWAY THE SHORTER LENGTH FROM THE LONGER.

$$3 - 2 = 1$$

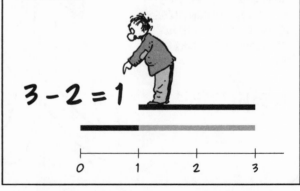

TO MAKE THIS PICTURE COVER ALL REAL NUMBERS, POSITIVE AND NEGATIVE, WE NEED TO THINK OF EACH NUMBER NOT AS A LENGTH, BUT AS AN **ARROW** WITH A **LENGTH AND DIRECTION.** ON THE NUMBER LINE, THIS ARROW POINTS FROM 0 TO THE NUMBER, SO NEGATIVE NUMBERS HAVE LEFT-POINTING ARROWS, WHILE POSITIVE ARROWS POINT RIGHT.

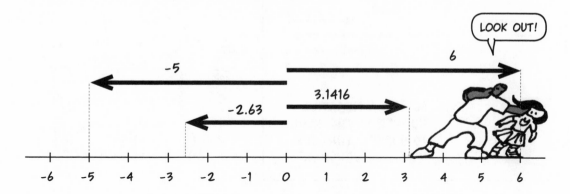

ADD TWO POSITIVE NUMBERS AS BEFORE: FIX ONE ARROW'S TAIL AT 0 AND MOVE THE OTHER'S TAIL TO THE FIXED ONE'S HEAD. THE SUM IS THE POSITION OF THE MOVED HEAD.

ADD NEGATIVE NUMBERS THE SAME WAY: HOLDING ONE ARROW FIXED, MOVE THE OTHER'S TAIL TO THE FIXED HEAD AND READ THE POSITION OF THE MOVED HEAD. TWO NEGATIVE NUMBERS, FOR EXAMPLE, ADD LIKE THIS:

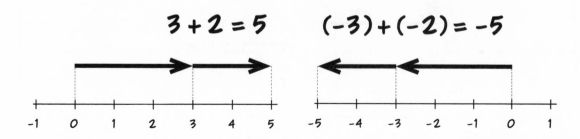

WHEN ADDING POSITIVE TO NEGATIVE, AGAIN PUT TAIL TO HEAD. THE SUM MAY BE POSITIVE...

OR NEGATIVE, DEPENDING ON THE NUMBERS BEING ADDED.

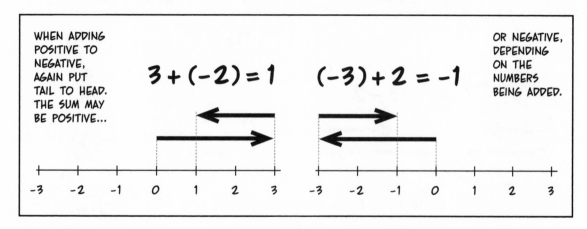

TAKE A CLOSER LOOK AT THE PICTURE ON P. 17 OF THE SUM 3 + (-2). IT'S VIRTUALLY THE SAME AS THE PICTURE OPPOSITE ON P. 16, OF THE DIFFERENCE 3 - 2. BOTH TAKE AWAY 2.

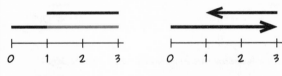

ADDING A NEGATIVE NUMBER IS THE SAME AS SUBTRACTING ITS "POSITIVE VERSION."

OKAYYY... THEN WHAT WOULD 3 + (-4) BE?

ER... 3 - 4? HOW CAN YOU TAKE 4 FROM 3?

READ ON...

THIS "POSITIVE VERSION" OF A NUMBER IS CALLED ITS **ABSOLUTE VALUE,** INDICATED BY SURROUNDING THE NUMBER WITH VERTICAL BARS, ||, AS IN |-2| = 2. THE ABSOLUTE VALUE IS A NUMBER'S (POSITIVE) SIZE, THE LENGTH OF ITS ARROW, ITS DISTANCE FROM 0. A POSITIVE NUMBER'S ABSOLUTE VALUE IS ITSELF, AND |0| = 0.

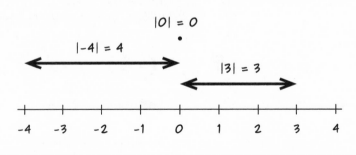

NOW LET'S LOOK AT (-3) + 2 AGAIN. ITS PICTURE IS THE NEGATIVE, OR MIRROR IMAGE, OF 3 + (-2) OR 3 - 2.

TO FIND THIS SUM, THEN, WE FIRST SUBTRACT 3 - 2 AND THEN **NEGATE** THE RESULT.

$$(-3) + 2 = -(3 - 2)$$
$$= -1$$

FIRST SUBTRACT, THEN NEGATE!

18

IN TERMS OF ABSOLUTE VALUE,
HERE ARE STEP-BY-STEP RULES
FOR ADDING ANY TWO NUMBERS,
WHETHER POSITIVE OR NEGATIVE:

POSITIVE + POSITIVE	NEGATIVE + NEGATIVE
ADD AS USUAL	ADD ABSOLUTE VALUES, THEN NEGATE
POSITIVE + NEGATIVE	

SUBTRACT ABSOLUTELY SMALLER FROM ABSOLUTELY
LARGER. THEN MAKE THE ANSWER'S SIGN THE SAME
AS THAT OF THE ABSOLUTELY LARGER NUMBER.

Example 1. FIND 4 + (−6).

4 IS POSITIVE AND −6 IS NEGATIVE, SO
WE SUBTRACT ABSOLUTE VALUES.

$$6 - 4 = 2$$

WE SEE THAT THE **NEGATIVE** NUMBER,
−6, HAS THE LARGER ABSOLUTE VALUE,
SO WE MAKE THE ANSWER NEGATIVE.

$$4 + (−6) = −2$$

Example 2. FIND (−2) + 9.

AGAIN WE SEE ONE POSITIVE AND ONE
NEGATIVE NUMBER, SO WE DO THE
SUBTRACTION.

$$9 - 2 = 7$$

THIS TIME, THOUGH, THE LARGER ABSOLUTE
VALUE BELONGS TO 9, THE **POSITIVE** NUM-
BER. SO WE LEAVE THE ANSWER POSITIVE.

$$(−2) + 9 = 7$$

THE LONGER ARROW
WINS THE BATTLE TO
CONTROL THE SIGN
OF THE ANSWER!

19

ANOTHER WAY TO THINK OF ADDING NEGATIVES IS IN TERMS OF **MONEY**... THIS IS HOW THE INDIAN MATHEMATICIAN **BHASKARA** THOUGHT ABOUT IT, WHEN HE MORE OR LESS INVENTED NEGATIVE NUMBERS ABOUT 1,500 YEARS AGO.

YOU CAN BLAME ME!

ASSETS, OR MONEY ON HAND PLUS MONEY OWED TO YOU, COUNT AS POSITIVE. **DEBTS,** MONEY YOU OWE TO OTHERS, COUNT AS NEGATIVE.

SO... ADD TWO ASSETS, GET A BIGGER ASSET.

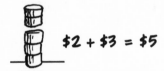

$$\$2 + \$3 = \$5$$

IF YOU OWE $2 TO FRED AND $3 TO FRIEDA, YOU OWE A TOTAL OF $5.

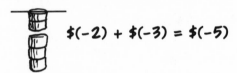

$$\$(-2) + \$(-3) = \$(-5)$$

IF YOU HAVE $3 IN ASSETS AND YOU OWE $2, YOU'RE STILL POSITIVE: YOU CAN PAY OFF YOUR DEBT AND STILL HAVE $1 LEFT.

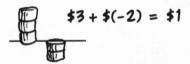

$$\$3 + \$(-2) = \$1$$

IF YOUR ASSETS TOTAL $2 AND YOU OWE $3, YOU ARE $1 SHORT OF BEING ABLE TO PAY. YOU "HAVE" NEGATIVE ONE DOLLAR.

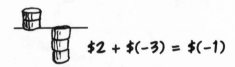

$$\$2 + \$(-3) = \$(-1)$$

THIS LEADS TO THE SAME ADDITION RULES AS BEFORE.

HOW COULD IT LEAD ANYWHERE ELSE?

Subtraction

SO FAR, WE'VE SEEN SUBTRACTIONS ONLY OF POSITIVE NUMBERS, AND THEN ONLY WHEN TAKING A SMALLER NUMBER FROM A LARGER ONE. BUT IF WE CAN ADD ANY TWO NUMBERS, WE SHOULD ALSO BE ABLE TO SUBTRACT ANY NUMBER FROM ANY OTHER. HERE'S HOW:

SUBTRACTING IS ADDING?

 **Subtracting a number is the same as adding its negative.**

THIS WAS TRUE WHEN SUBTRACTING A POSITIVE NUMBER FROM A LARGER POSITIVE NUMBER: $5 - 3 = 5 + (-3)$. NOW WE SIMPLY **DEFINE** SUBTRACTION FOR OTHER NUMBERS TO WORK IN THE SAME WAY. FOR EXAMPLE:

$$2 - 3 = 2 + (-3) = -1$$
$$-6 - 7 = -6 + (-7) = -13$$

NOTE ESPECIALLY: SUBTRACTING A NEGATIVE NUMBER MEANS ADDING **ITS** NEGATIVE, WHICH IS **POSITIVE.**

SUBTRACTING A DEBT MAKES YOU RICHER!

$$9 - (-3) = 9 + 3 = 12$$

REMEMBER, $-(-3) = 3$!

$$-6 - (-2) = -6 + 2 = -4$$

AND WITH THAT, YOU SHOULD BE READY TO SOLVE SOME PRACTICE PROBLEMS YOURSELF!

Problems

1. DO THE SUMS.

a. $(-4) + 8$

b. $(-3) + (-5)$

c. $9 + (-3)$

d. $|-14.5| + (-15.6)$

e. $\dfrac{5}{2} + (-2)$

f. $\left(-\dfrac{1}{2}\right) + \dfrac{1}{3}$

2. SUBTRACT.

a. $10 - (-9)$

b. $9 - (-10)$

c. $(-9) - 10$

NOTE THAT IN PROBLEM 2c, WE COULD LEAVE OUT THE PARENTHESES AND SIMPLY WRITE $-9 - 10$.

d. $-4 - 8$

e. $4 - 8$

f. $|-4| - 6$

g. $\dfrac{9}{16} - \dfrac{7}{12}$

h. $6 - |2|$

i. $|2 - 100|$

3. WHAT IS $-5 + 3 - 6 + 4 + (-2)$?

4. ARE THE SUMS OF THESE PAIRS OF ARROWS POSITIVE OR NEGATIVE?

a.   b.   c.   d.

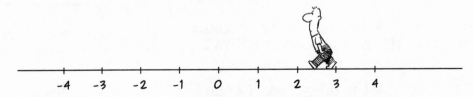

5. SUPPOSE YOU'RE TAKING A WALK ON THE NUMBER LINE. IF YOU START AT 3, THEN WALK 6 UNITS IN A NEGATIVE DIRECTION, THEN REVERSE COURSE AND WALK 2 UNITS IN A POSITIVE DIRECTION, WHERE WOULD YOU END UP? WHERE WOULD THE SAME WALK END IF YOU HAD STARTED AT −200 INSTEAD OF 3?

6. BOYCE HAS $5 IN HIS POCKET. HE BORROWS $10 FROM HIS FRIEND FRANCINE. THEN HE LOSES $8 ON A STUPID BET ABOUT THE OUTCOME OF A SCHOOL ELECTION. WHAT IS BOYCE'S NET FINANCIAL POSITION AT THE END?

7. JESSICA OWES $5 TO ANGELA AND $2 TO BARBARA. JESSICA HAS $20 ON HAND.

a. WHAT IS JESSICA'S NET WORTH (THE SUM OF EVERYTHING, COUNTING DEBTS AS NEGATIVE).

b. NOW ANGELA "FORGIVES" $3 OF JESSICA'S DEBT, I.E., CANCELS IT SO THAT JESSICA NO LONGER HAS TO PAY THE $3. WRITE THIS AS A SUBTRACTION OF A NEGATIVE NUMBER.

c. WHAT IS JESSICA'S NET FINANCIAL POSITION AT THE END?

Chapter 3
Multiplication and Division

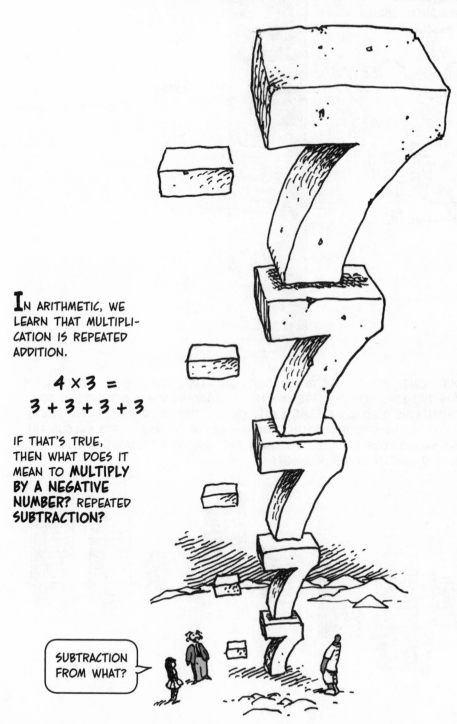

In arithmetic, we learn that multiplication is repeated addition.

$$4 \times 3 =$$
$$3 + 3 + 3 + 3$$

If that's true, then what does it mean to **MULTIPLY BY A NEGATIVE NUMBER?** Repeated **SUBTRACTION?**

SUBTRACTION FROM WHAT?

TO SEE HOW THIS WORKS, WE STAY WITH BHASKARA A BIT LONGER AND THINK IN TERMS OF MONEY. POSITIVE MONEY SITS ABOVE A HORIZONTAL LINE, NEGATIVE MONEY BELOW IT.

MY FAVORITE PART!

ASSETS

0

LIABILITIES

IN REAL LIFE, YOUR MONEY MAY CHANGE DAY BY DAY... AND **TIME** CAN ALSO BE POSITIVE OR NEGATIVE. TODAY IS THE ZERO POINT; YESTERDAY IS −1; TOMORROW +1; AND SO ON, SO THAT THE HORIZONTAL LINE BECOMES A **TIME LINE.** ON ANY DAY, YOUR ASSETS AND DEBTS APPEAR AS A STACK OF COINS STRADDLING THE LINE, ASSETS ABOVE AND DEBTS BELOW. THE STACK AT EACH DAY SHOWS YOUR FINANCIAL POSITION THAT DAY. FOR INSTANCE, ON DAY 4 YOU OWE 3 COINS AND HAVE 14 COINS IN ASSETS.

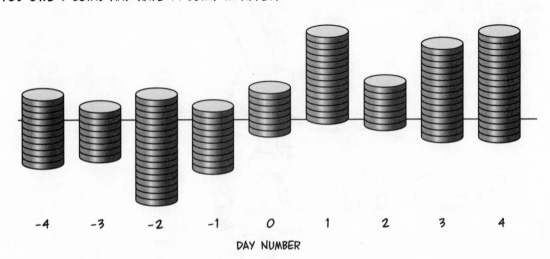

| −4 | −3 | −2 | −1 | 0 | 1 | 2 | 3 | 4 |

DAY NUMBER

NOW LET'S MULTIPLY MONEY BY TIME. SUPPOSE CELIA HAS BEEN BETTING $2 EVERY DAY FOR A LONG TIME (BETTING BORROWED MONEY IF SHE'S "IN THE HOLE," I.E., BELOW ZERO). AND SUPPOSE THAT TODAY, AT TIME 0, SHE HAS $0.

PLUS × PLUS

IF CELIA WINS $2 EVERY DAY FROM NOW ON, THEN ON DAY 3 SHE WILL HAVE $6.

$$3 \times 2 = 6$$

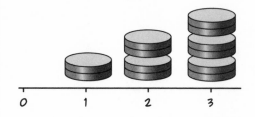

MINUS × PLUS

IF SHE HAS BEEN WINNING $2 DAILY, THEN THREE DAYS AGO, ON DAY −3, SHE MUST HAVE HAD $(−6) TO REACH $0 TODAY.

$$(-3) \times 2 = -6$$

WINNING MAKES DEBT SMALLER.

PLUS × MINUS

IF SHE LOSES $2 DAILY, ON DAY 3 SHE WILL HAVE $(−6)

$$3 \times (-2) = -6$$

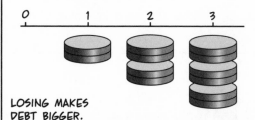

LOSING MAKES DEBT BIGGER.

MINUS × MINUS

IF SHE HAS BEEN LOSING $2 DAILY, ON DAY −3 SHE HAD $6.

$$(-3) \times (-2) = 6$$

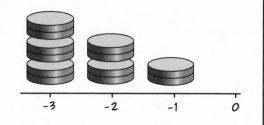

LOAN SHARK →

COME BACK LAST WEEK. I CAN PAY YOU THEN!

TO SUMMARIZE, HERE IS A LITTLE TABLE SHOWING THE **SIGN RULE FOR MULTIPLYING POSITIVE AND NEGATIVE NUMBERS:**

	+	−
+	+	−
−	−	+

OR, IF YOU PREFER IT WRITTEN IN WORDS...

POSITIVE · POSITIVE = POSITIVE
NEGATIVE · POSITIVE = NEGATIVE
POSITIVE · NEGATIVE = NEGATIVE
NEGATIVE · NEGATIVE = POSITIVE

Examples: $5 \times (-2) = -10$, $(-3)(-7) = 21$, $(-4) \times 4 = -16$

ANOTHER WAY TO PUT IT: MULTIPLYING BY A POSITIVE NUMBER LEAVES THE OTHER NUMBER'S SIGN UNCHANGED. MULTIPLYING BY A NEGATIVE NUMBER REVERSES THE SIGN.

MULTIPLYING BY 6 GIVES AN ANSWER WITH THE **SAME** SIGN AS −2; MULTIPLYING BY −2 GIVES AN ANSWER WITH SIGN **OPPOSITE** FROM THAT OF 6.

WHEN ONE OF THE NEGATIVE NUMBERS IS −1, THE RULE SAYS: MULTIPLY BY 1 AND CHANGE THE SIGN OF THE OTHER FACTOR. **MULTIPLYING BY −1 IS THE SAME AS TAKING A NUMBER'S NEGATIVE.**

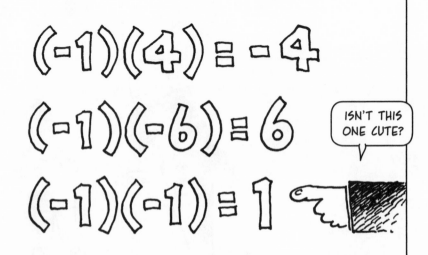

ISN'T THIS ONE CUTE?

Multiplication Without Money

HERE ARE THREE ROWS OF TWO SQUARES EACH. THAT'S THREE TWOS, OR 3 × 2. THE **PRODUCT** OF TWO NUMBERS (THE RESULT OF MULTIPLYING THEM TOGETHER) LOOKS LIKE A **RECTANGLE**: EACH SIDE IS ONE OF THE NUMBERS.

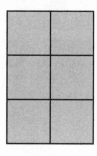

EACH SMALL SQUARE IS 1 UNIT ON EACH SIDE, AND THE GRAY RECTANGLE'S AREA IS **THE NUMBER OF UNIT SQUARES IT CONTAINS.** THE UNIT SQUARE, WHICH CONTAINS EXACTLY ONE OF ITSELF, HAS AREA = 1.

THIS WORKS EVEN IF THE SIDES AREN'T WHOLE NUMBERS. HERE THE GRAY RECTANGLE HAS SIDES OF LENGTH $\frac{1}{2}$ AND $\frac{1}{3}$. (WE'VE ENLARGED THE UNIT SQUARE HERE.)

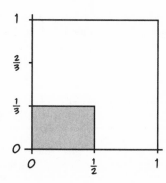

YOU CAN SEE THAT SIX OF THE GRAY RECTANGLES FIT TOGETHER EXACTLY TO MAKE A UNIT SQUARE, SO ONE GRAY RECTANGLE'S AREA IS $\frac{1}{6}$, THE PRODUCT OF $\frac{1}{3}$ AND $\frac{1}{2}$.

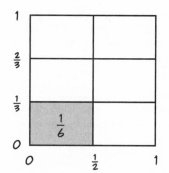

$$\frac{1}{3} \cdot \frac{1}{2} = \frac{1}{6}$$

HERE IS THE PRODUCT OF MORE COMPLICATED FRACTIONS, (5/3)·(5/2). THE UNIT SQUARE IS OUTLINED IN BLACK.

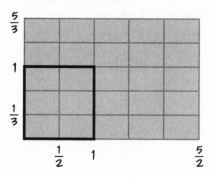

$$\frac{5}{3} \cdot \frac{5}{2} = \frac{25}{6}$$

EACH SMALL RECTANGLE IS 1/6, AND THERE ARE 5×5 = 25 OF THEM.

THE PICTURE IS GOOD NO MATTER WHAT THE SIDES: A RECTANGLE'S AREA IS THE PRODUCT OF THE LENGTHS OF THE TWO SIDES.

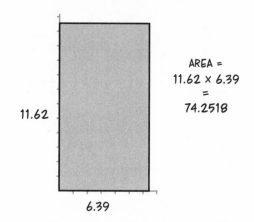

AREA =
11.62 × 6.39
=
74.2518

WE COULD GO ON HERE AND DRAW RECTANGLES WITH NEGATIVE SIDES, BUT IT'S REALLY NOT WORTH IT.

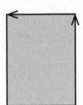

INSTEAD I'D LIKE TO SHOW YOU ANOTHER PICTURE OF MULTI-PLICATION THAT YOU WON'T SEE IN REGULAR ALGEBRA TEXTBOOKS...

LET IT BE OUR LITTLE SECRET!

THIS PICTURE SHOWS MULTIPLICATION IN TERMS OF "SCALING," WHICH IS LIKE ENLARGING OR REDUCING A PHOTOGRAPH. ONLY INSTEAD OF A PHOTO, WE'LL CHANGE THE SCALE OF THE **ENTIRE NUMBER LINE.**

IMAGINE TWO NUMBER LINES, ONE OF WHICH IS STRETCHED OUTWARD FROM 0 UNTIL ALL LENGTHS ARE DOUBLED. HERE IT'S THE UPPER.

TO FIND THE PRODUCT OF **2** TIMES ANY NUMBER, JUST LOOK **DIRECTLY BELOW THAT NUMBER.**

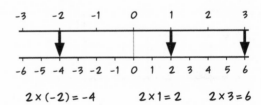

$$2 \times (-2) = -4 \qquad 2 \times 1 = 2 \qquad 2 \times 3 = 6$$

THIS IS COOL, BECAUSE INSTEAD OF SHOWING A SINGLE PRODUCT, LIKE 2×3, THE PICTURE SHOWS YOU THE PRODUCT OF 2 TIMES **EVERYTHING!**

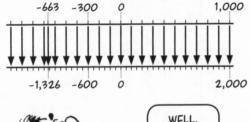

WE CAN ALSO SCALE THE LINE **DOWN** TO MULTIPLY BY A NUMBER BETWEEN 0 AND 1. MULTIPLICATION BY 1/2 **SQUEEZES** THE LINE UNTIL ALL LENGTHS ARE HALVED:

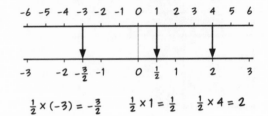

$$\tfrac{1}{2} \times (-3) = -\tfrac{3}{2} \qquad \tfrac{1}{2} \times 1 = \tfrac{1}{2} \qquad \tfrac{1}{2} \times 4 = 2$$

YOU'LL HAVE AN OPPORTUNITY TO PLAY WITH THIS PICTURE IN THE PROBLEM SET AT THE END OF THE CHAPTER.

WELL, I THINK IT'S COOL, ANYWAY!

Division

DIVIDING SOMETHING BY A POSITIVE WHOLE NUMBER MEANS BREAKING THE THING INTO SO MANY EQUAL PARTS AND MEASURING THE SIZE OF ONE PART. HERE, FOR INSTANCE, IS A PICTURE OF $6 \div 2$.

THE SIX UNITS ARE SPLIT OR, YES, DIVIDED, INTO TWO EQUAL GROUPS, AND WE SEE THAT EACH GROUP IS THREE UNITS. $6 \div 2 = 3$

HERE'S ANOTHER WAY OF SEEING THE SAME THING. NO MATTER WHICH WAY YOU CUT IT IN TWO EQUAL PARTS, THE STUFF IN ANY ONE PART AMOUNTS TO 3.

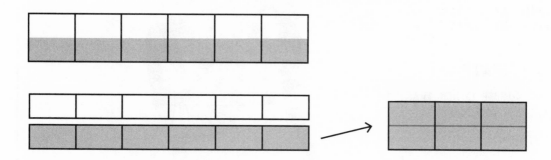

THE PICTURE ALSO SHOWS THAT DIVIDING BY 2 IS THE SAME AS MULTIPLYING BY $\frac{1}{2}$.

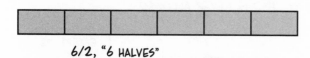

6/2, "6 HALVES"

FOR THIS REASON, WE RARELY WRITE THE DIVISION SYMBOL $\div$ IN ALGEBRA. INSTEAD, WE USE A FRACTION BAR, EITHER HORIZONTAL (——) OR TILTED (/). IT'S ALL THE SAME!

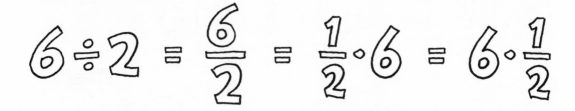

$$6 \div 2 = \frac{6}{2} = \frac{1}{2} \cdot 6 = 6 \cdot \frac{1}{2}$$

THE NUMBERS 2 AND $\frac{1}{2}$ ARE SAID TO BE **RECIPROCAL** TO EACH OTHER. THIS MEANS THAT THEIR PRODUCT EQUALS 1. ANY PAIR OF NUMBERS WHOSE PRODUCT IS 1 ARE CALLED EACH OTHER'S RECIPROCAL. 6 AND 1/6, 1,000 AND 1/1,000, 32.642 AND 1/(32.642).

YOU'RE MINE!

AND YOU'RE MINE!

KIND OF SWEET, ISN'T IT?

$$\frac{1}{1,000} \times 1,000 = 1$$

GIVEN ANY NUMBER OTHER THAN ZERO, YOU CAN FIND ITS RECIPROCAL BY LONG DIVISION OR ON A CALCULATOR. ONLY ZERO HAS NO RECIPROCAL, BECAUSE 0 × ANYTHING = 0. THERE'S **NO NUMBER** THAT MAKES THIS EQUATION WORK:

$$0 \times \text{WHAT?} = 1$$

EVERY NUMBER OTHER THAN 0 HAS A RECIPROCAL.

1/(32.642) = 0.030635377734207...

(0.030635377734207...) × (32.642) = 1

AS FAR AS WE'RE CONCERNED, **DIVISION BY ANY NUMBER IS THE SAME AS MULTIPLICATION BY ITS RECIPROCAL.** DIVISION BY ZERO IS NEVER ALLOWED.

$$7 \div 5 = 7 \times \frac{1}{5}$$

SAD... I'VE SUNK BELOW THE FRACTION BAR... OH, WELL... COULD HAPPEN TO ANYONE, I GUESS...

EXCEPT ME, SUCKA!

WE SAW THAT $\frac{1}{2}$ HAS THE RECIPROCAL 2, OR 2/1, WHICH IS $\frac{1}{2}$ TURNED UPSIDE DOWN. IN THE SAME WAY, YOU CAN FIND THE RECIPROCAL OF **ANY** FRACTION SIMPLY BY TURNING IT OVER.

TURNING OVER, OR **INVERTING,** A FRACTION MEANS EXCHANGING ITS TOP—THE **NUMERATOR**—WITH ITS BOTTOM—THE **DENOMINATOR**—TO MAKE A NEW FRACTION CALLED ITS **INVERSE**. $\frac{2}{3}$ BECOMES $\frac{3}{2}$. TURNING IT OVER **TWICE,** OF COURSE, RESTORES THE ORIGINAL FRACTION, SO REALLY THE PAIR ARE INVERSE TO EACH OTHER.

$$\frac{27}{15} \longleftrightarrow \frac{15}{27}$$

A PAIR OF INVERSE FRACTIONS.

NOW ANY PAIR OF INVERSES ARE, IN FACT, EACH OTHER'S RECIPROCALS. YOU CAN SEE THIS BY MULTIPLYING THEM TOGETHER. THE PRODUCT'S NUMERATOR AND DENOMINATOR ARE EQUAL, SO THE PRODUCT = 1.

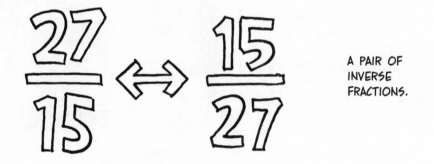

$$\frac{3}{2} \cdot \frac{2}{3} = \frac{3 \times 2}{2 \times 3} = \frac{6}{6} = 1$$

NOW WE CAN MAKE SOME SENSE OF THAT STRANGE RULE FOR DIVIDING BY A FRACTION: "INVERT AND MULTIPLY." FOR US, DIVISION **MEANS** MULTIPLYING BY THE RECIPROCAL, AND A FRACTION'S RECIPROCAL IS ITS INVERSE.

$$3 \div \frac{2}{5} \quad \left(\text{OR } \frac{3}{\left(\frac{2}{5}\right)} \right) \text{ MEANS}$$

$$3 \times \left(\text{RECIPROCAL OF } \frac{2}{5} \right) \text{ OR}$$

$$3 \times \frac{5}{2} = \frac{15}{2}$$

THIS HANDY RULE SAVES US FROM TRYING TO UNDERSTAND DIVISION IN TERMS OF SPLITTING SOMETHING UP. THAT'S FINE FOR DIVIDING BY POSITIVE WHOLE NUMBERS, BUT HOW WOULD YOU "DIVIDE" SOMETHING INTO, SAY, 54/17 EQUAL PARTS?

ER... UM...

ANSWER: **DON'T WORRY ABOUT IT!** SIMPLY MULTIPLY BY THE RECIPROCAL.

BY THE WAY, WHY DID YOU INVITE 54/17 PEOPLE TO THIS PARTY IN THE FIRST PLACE?

SOME OF THE PLATES WERE BROKEN...

IN THE PROBLEMS AT THE END OF THIS CHAPTER, I'LL SHOW YOU ANOTHER WAY TO THINK OF DIVIDING BY A FRACTION.

Negative Fractions and Reciprocals

ON P. 26, WE SAW THAT MULTIPLICATION BY –1 NEGATES ANY NUMBER, AND THAT INCLUDES –1 ITSELF: $(-1)(-1) = 1$. IN OTHER WORDS, –1 IS ITS OWN RECIPROCAL!

$$\frac{1}{-1} = -1$$

NOW LET'S TRY DIVIDING ANY OLD POSITIVE NUMBER BY A NEGATIVE NUMBER, SAY $3/(-4)$.

$$\frac{3}{-4} = \frac{1 \times 3}{-1 \times 4}$$

$$= \frac{1}{-1} \times \frac{3}{4}$$

$$= (-1)\frac{3}{4}$$

$$= -\frac{3}{4}$$

LOCATING $-\frac{3}{4}$ ON THE NUMBER LINE, WE SEE IT'S AN ORDINARY NEGATIVE NUMBER.

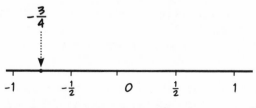

IT'S ALSO EASY TO SHOW THAT $(-3)/4 = -\frac{3}{4}$.

THIS SHOWS THAT NEGATIVE DIVIDED BY POSITIVE, OR VICE VERSA, IS NEGATIVE. IT'S ALSO TRUE THAT NEGATIVE ÷ NEGATIVE IS POSITIVE, BECAUSE

$$\frac{-2}{-7} = \frac{(-1) \times 2}{(-1) \times 7} = \left(\frac{-1}{-1}\right)\left(\frac{2}{7}\right) = \frac{2}{7}$$

ANYTHING OVER ITSELF IS 1.

A POSITIVE NUMBER

IN OTHER WORDS, THE SIGN RULES FOR DIVISION ARE THE SAME AS FOR MULTIPLICATION.

$$\frac{\text{POSITIVE}}{\text{NEGATIVE}} = \frac{\text{NEGATIVE}}{\text{POSITIVE}} = \text{NEGATIVE}$$

$$\frac{\text{NEGATIVE}}{\text{NEGATIVE}} = \frac{\text{POSITIVE}}{\text{POSITIVE}} = \text{POSITIVE}$$

IN PARTICULAR, THE RECIPROCAL OF A NEGATIVE NUMBER MUST BE NEGATIVE, AND A NEGATIVE FRACTION'S RECIPROCAL IS ITS INVERSE, STILL WITH THE MINUS SIGN ATTACHED.

$$\left(-\frac{3}{4}\right)\left(-\frac{4}{3}\right) = \frac{(-3)\cdot(-4)}{4 \times 3}$$

$$= \frac{12}{12}$$

$$= 1$$

 NOW LET'S SOLVE SOME PROBLEMS!

Problems

1. MULTIPLY.

a. $9 \times (-3)$

b. $(-2)(-2)$

c. $(-2)(-3)(-4)$

d. $\left(\frac{2}{3}\right)\left(-\frac{3}{4}\right)$

e. $\left(-\frac{1}{2}\right)(50)$

f. $\left(-\frac{1}{2}\right)\left(-\frac{1}{2}\right)$

g. $(-1)(6+3)$
(REMEMBER: DO THE SUM INSIDE THE PARENTHESES FIRST.)

h. $(-1)(2-4)$

i. $0 \times (-0.3569)$

2. DIVIDE.

a. $15 / (-3)$

b. $\dfrac{-20}{-4}$

c. $\dfrac{0}{-5}$

d. $\dfrac{-3{,}507.89}{1}$

3. WHAT IS THE RECIPROCAL OF -2? OF $-\frac{1}{3}$? DOES 0 HAVE A RECIPROCAL?

4. WHAT IS $\left(\frac{3}{2}\right)\left(\frac{2}{3}\right)(50)$?

5. $\left(\frac{7}{8}\right)\left(\frac{8}{7}\right)(-31) = ?$

6. HERE ARE TWO NUMBER LINES CENTERED AT 0, THE UPPER LINE BEING SCALED UP BY A FACTOR OF 3.

a. WHAT NUMBER ON THE LOWER LINE IS BELOW THE UPPER LINE'S 1?

b. ON THE UPPER LINE, $\frac{1}{3}$ IS ABOVE WHAT ON THE LOWER LINE?

7. DRAW A PICTURE OF AN UPPER LINE SCALED BY 2/3 AND AN UNSCALED LINE BELOW WITH THE ZEROES LINED UP. WHERE ON THE UPPER LINE IS 3/2?

8. IF YOU SCALE THE UPPER LINE BY ANY NUMBER, WHERE IS THAT NUMBER ON THE LOWER LINE, RELATIVE TO THE UPPER LINE? WHERE IS THE NUMBER'S RECIPROCAL ON THE UPPER LINE?

9. WHAT DO YOU SUPPOSE THIS PICTURE MIGHT LOOK LIKE WHEN MULTIPLYING BY -1?

10. SUPPOSE WE HAVE A CAKE AND $2\frac{1}{2}$ PLATES. WHAT HAPPENS WHEN WE DIVIDE THE CAKE BY $2\frac{1}{2}$?

WELL, $2\frac{1}{2} = 5/2$, SO, BLINDLY FOLLOWING THE RULES, WE INVERT THIS TO 2/5 AND MULTIPLY BY OUR ONE CAKE.

$$1 \times \frac{2}{5} = \frac{2}{5}$$

WE GET AN ANSWER OF 2/5 CAKE. TO SERVE THE CAKE, WE CUT IT INTO FIFTHS.

WE CAN THEN PUT 2 FIFTHS ON EACH WHOLE PLATE, AND THE REMAINING 1 ON THE HALF PLATE.

VOILÀ! THE CAKE IS IN FACT DIVIDED INTO $2\frac{1}{2}$ PARTS! A "PART," OR $1 \div 2\frac{1}{2} = 2/5$, IS THE AMOUNT THAT ENDS UP ON EACH WHOLE PLATE. HALF A PART GOES ON THE HALF-PLATE.

NOW SUPPOSE WE HAD $2\frac{1}{3}$ PLATES. DOES A SIMILAR TRICK WORK? (NOW WE ARE DIVIDING BY 7/3.) HOW ABOUT $2\frac{2}{3}$? $10\frac{3}{4}$?

Chapter 4
Expressions and Variables

In MATH, THE ACT OF DOING AN ADDITION, SUBTRACTION, MULTIPLICATION, OR THE LIKE IS KNOWN AS "PERFORMING AN OPERATION," AS IF THE POOR NUMBERS WERE HAVING SURGERY.

IN THIS CHAPTER, WE PUT MULTIPLE OPERATIONS TOGETHER TO FORM **EXPRESSIONS**... AND THESE EXPRESSIONS WILL INCLUDE NOT ONLY NUMBERS BUT ALSO LETTERS OR "VARIABLES," WHATEVER THAT MEANS. BY CHAPTER'S END, YOU'LL BE OPERATING ON THINGS THAT LOOK LIKE THIS:

WE'LL TRY TO KEEP THE BLEEDING TO A MINIMUM.

INSTEAD OF CUTTING UP BODIES, LET'S START BY BUILDING A BOOKCASE. IT WILL HAVE 5 SHELVES, AND EACH SHELF WILL BE 3 FEET LONG. THE SHELVES' TOTAL LENGTH IS OBVIOUSLY

 FEET

(I KNOW, I KNOW, THAT'S 15 FEET, BUT WE DON'T CARE ABOUT THAT AT THE MOMENT...)

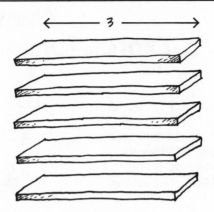

IF WE ADD TWO 4-FOOT SIDES, THE AMOUNT OF LUMBER INCREASES... AND WE HAVE TO ADD THIS MUCH:

 FEET

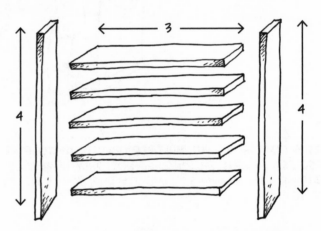

SO THIS **NUMERICAL EXPRESSION,** THE SUM OF THE TWO PRODUCTS, GIVES THE TOTAL LENGTH OF ALL THE BOARDS:

$$(5 \times 3) + (2 \times 4)$$

FOUR NUMBERS, SEVERAL OPERATIONS!

WHO KNEW YOU NEEDED A SURGEON TO BUILD A BOOK-CASE?

YOU KNOW WHAT THE PARENTHESES MEAN: DO THE OPERATION **INSIDE** THE PARENTHESES—THE MULTIPLICATIONS IN THIS CASE—**BEFORE** DOING THE ADDITION OUTSIDE. DOING THE ARITHMETIC GIVES THE EXPRESSION'S **VALUE**.

FIRST THE "INSIDES":

$$5 \times 3 = 15$$
$$2 \times 4 = 8$$

THEN THE ADDITION:

$$15 + 8 = 23$$

THE VALUE

THE PLACEMENT OF PARENTHESES MATTERS. WE GET A DIFFERENT VALUE IF OPERATIONS GO IN A DIFFERENT ORDER:

$$(5 \times 3) + (2 \times 4) = 15 + 8 = 23$$
$$5 \times (3 + 2) \times 4 = 5 \times 5 \times 4 = 100$$

SAME NUMBERS, SAME OPERATIONS, DIFFERENT ORDER!

IN THIS WAY, MATH IS LIKE THE REST OF THE WORLD: OUTCOMES OFTEN DEPEND ON WHAT GOES FIRST.

1. SET DOWN GLASS.
2. POUR MILK.

1. POUR MILK.
2. SET DOWN GLASS.

HOW MANY TIMES DO I HAVE TO SAY IT: FIRST CUT, THEN STITCH!!

ALTHOUGH ORDER CERTAINLY MATTERS, IT'S ALSO TRUE THAT TOO MANY SETS OF PARENTHESES CAN REALLY JUNK UP AN EXPRESSION.

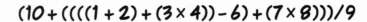

$$(10 + ((((1 + 2) + (3 \times 4)) - 6) + (7 \times 8)))/9$$

NOW THAT SHOULD BE UNCONSTITUTIONAL!

CRUEL **AND** UNUSUAL...

THE IDEA IS TO BE CLEAR WITH AS FEW PARENTHESES AS POSSIBLE... SO THE MATH WORLD HAS AGREED ON A WAY TO SHED THEM. CALL IT THE **ORDER OF OPERATIONS RULE:**

ORDER! ORDER!

If no parentheses are present, multiply and divide before adding and subtracting.

FOLLOWING THIS RULE, THE BOOKSHELF EXPRESSION IS

$$5 \cdot 3 + 2 \cdot 4$$

AND WE'RE IN NO DANGER OF GETTING IT WRONG. MULTIPLICATION COMES FIRST.

Examples:

1. EVALUATE (FIND THE VALUE OF) 1 − 2·3

SOLUTION: NO PARENTHESES ARE PRESENT, SO DO THE MULTIPLICATION FIRST: 2·3 = 6. THEN SUBTRACT. 1 − 6 = −5

2. EVALUATE $1 - \dfrac{4}{-2}$.

SOLUTION: DIVISION COMES FIRST. 4/(−2) = −2. THEN SUBTRACT. 1 − (−2) = 3.

3. EVALUATE 3(4/6 + 2·7).

SOLUTION: WHEN PARENTHESES ARE PRESENT, WE MUST EVALUATE THE INSIDE EXPRESSION FIRST! THAT EXPRESSION HAS BOTH ADDITION AND MULTIPLICATION/DIVISION. WE DO THE MULTIPLICATION AND DIVISION FIRST. 4/6 = 2/3 AND 2·7 = 14. NEXT ADD: $14 + \frac{2}{3}$ = 44/3. NOW THAT THE INSIDE QUANTITY HAS BEEN FOUND, MULTIPLY IT BY 3.

3(44/3) = 44

NOW LET'S GET

ALGEBRAIC!

THIS PAGE, READER, MARKS THE PLACE
WHERE WE CROSS FROM THE OLD,
FAMILIAR GROUND OF ARITHMETIC TO
THE PROMISED LAND OF ALGEBRA.

THE CHANGE BEGINS WITH A QUESTION ABOUT OUR BOOKCASE: CAN WE WRITE AN
EXPRESSION FOR THE TOTAL LENGTH OF ALL BOARDS OF A 4-FOOT-TALL BOOKCASE
WITH 5 SHELVES **OF ANY LENGTH?**

OF COURSE WE CAN! IF WE WRITE "LENGTH" FOR
THE LENGTH OF A SINGLE SHELF—WHATEVER IT
IS—THEN THE 5-SHELF UNIT, INCLUDING ITS SIDES,
HAS A TOTAL BOARD LENGTH OF

$$5 * \text{LENGTH} + 2 * 4$$

IT'S NOT A NUMBER, BUT
RATHER A FORMULA FOR FINDING
A NUMBER, GIVEN ANY SHELF LENGTH.

THE WORD "LENGTH" IN THE EXPRESSION 5 × LENGTH + 2 × 4 IS CALLED A **VARIABLE**, BECAUSE IT STANDS IN FOR ALL THE **VARIOUS** LENGTHS THAT A SHELF MIGHT HAVE.

LIKE 3 OR 3.1 OR 3.12657 OR 3.12658 OR 9.10104 OR...

OKAY, OKAY! WE GET IT!

WE COULD ALSO VARY THE **HEIGHT** OF THE BOOKCASE, RATHER THAN KEEPING IT 4 FEET. THEN THE TOTAL BOARD LENGTH HAS THIS EXPRESSION:

$$5 \times LENGTH + 2 \times HEIGHT$$

THE **NUMBER OF SHELVES** COULD ALSO VARY. WRITING "NUMBER" FOR THE NUMBER OF SHELVES GIVES THIS EXPRESSION

$$NUMBER \times LENGTH + 2 \times HEIGHT$$

FOR THE TOTAL LENGTH OF ALL THE BOARDS.

THE WORDS "NUMBER," "LENGTH," AND "HEIGHT" ARE ALL VARIABLES. AN EXPRESSION IS CALLED **ALGEBRAIC** IF IT CONTAINS ONE OR MORE VARIABLES.

DUDE, WHERE ARE THE BOOKS?

IT'S AN ALGEBRAIC CASE—IT CONTAINS VARIABLES!

DISTANCE TIME WEIGHT age Speed Price TAX RATE AT-BATS HITS

THE VARIABLE NAMES
WE'VE JUST SEEN, LIKE
"LENGTH," ARE ENTIRE
WORDS, AND IN SOME
FIELDS PEOPLE WRITE
VARIABLES OUT IN FULL
THAT WAY. COMPUTER
PROGRAMMERS, FOR
INSTANCE, ADORE LONG
VARIABLE NAMES FOR
REASONS OF THEIR OWN.
HERE'S A SAMPLE.

```
PROCEDURE ReadSchedClrArgs(

    VAR  StartDay, EndDay: DayType;
    VAR  StartHour, EndHour: HourType;
    VAR  Error: boolean);
    VAR  InputHour: integer;

    FUNCTION MapTo24(Hour: integer): HourType;
        CONST
                    { AM/PM time cut-off. }
    LastPM = 5;
        BEGIN
    IF Hour <=  LastPM THEN
      MapTo24 := Hour + 12
    ELSE
       MapTo24 := Hour
             END;
```

IN ALGEBRA, THOUGH, WE NEARLY ALWAYS ABBREVIATE VARIABLES TO **SINGLE LETTERS.**
THAT'S BECAUSE WE'LL HAVE TO WRITE OUR VARIABLES OVER AND OVER AS WE PUSH
EXPRESSIONS AROUND. WE WANT SOMETHING SHORT. ALGEBRA IS LIKE TEXTING!

ARE YOU
SURE YOU'RE
NOT JUST
BEING LAZY?

LOL! IT'S
OK 2 B
LAZY!

HERE'S WHAT HAPPENS TO
OUR BOOKCASE EXPRESSIONS.
NOTE ALSO THAT ALL MULTI-
PLICATION SIGNS HAVE COM-
PLETELY VANISHED ALONG
WITH THE EXTRA LETTERS. IN
ALGEBRA, **MULTIPLICATION
IS SHOWN SIMPLY BY
PUTTING TWO FACTORS
SIDE BY SIDE.**

$$5L + 8$$
$$5L + 2H$$
$$nL + 2H$$

SBSM!*

*SHORT
BUT STILL
MEANINGFUL

41

More Examples

IF YOU MOVE AT A STEADY SPEED OF 60 MILES PER HOUR, THEN IN t HOURS YOU TRAVEL A DISTANCE OF

$60t$ MILES

A RECTANGLE'S **AREA** IS THE PRODUCT OF THE HEIGHT h AND ITS WIDTH w. AREA = hw.

ITS **PERIMETER** IS THE TOTAL DISTANCE AROUND THE RECTANGLE, THE SUM OF ALL ITS SIDES. YOU COULD WRITE THIS AS EITHER

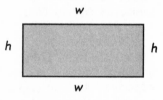

$2h + 2w$ (DOUBLE EACH SIDE, THEN ADD) OR

$2(h + w)$ (ADD HEIGHT TO WIDTH, THEN DOUBLE)

WHERE I LIVE, **THE SALES TAX RATE IS 8%** (THAT'S 8/100 = .08). IF AN ITEM IS MARKED WITH A PRICE p, THEN THE SALES TAX IS .08p. THE PRICE I ACTUALLY PAY IS THE MARKED PRICE PLUS THE TAX, OR

$p + .08p$

IF YOU PLAN TO WALK 100 MILES, AND YOU'VE ALREADY COVERED x OF THEM, THEN THE DISTANCE STILL TO GO IS

$100 - x$

ALGEBRA: YOU CAN'T WALK AWAY FROM IT!

YOU READ ALGEBRAIC EXPRESSIONS MUCH AS YOU READ TEXT MESSAGES: ONE LETTER, NUMBER, OR SYMBOL AT A TIME—EXCEPT THAT PARENTHESES INDICATE GROUPING. WHATEVER IS INSIDE PARENTHESES IS CALLED A "QUANTITY."

EXPRESSION	HOW TO SAY IT	MEANING
$a + x$	"AY PLUS EX"	THE SUM OF TWO NUMBERS
$5y$	"FIVE WYE"	FIVE TIMES A NUMBER
$\dfrac{x}{2}$	"EX OVER TWO"	HALF A NUMBER
$-a$	"NEGATIVE AY"	THE NEGATIVE OF A NUMBER
$5T + 1$	"FIVE TEE PLUS ONE"	ONE MORE THAN FIVE TIMES A NUMBER
$5(x + 1)$	"FIVE TIMES THE QUANTITY EX PLUS ONE"	FIVE TIMES THE SUM OF A NUMBER AND ONE

EVALUATING EXPRESSIONS

UNLIKE A NUMERIC EXPRESSION, AN ALGEBRAIC EXPRESSION HAS NO DEFINITE VALUE: $5L + 8$ ISN'T A NUMBER. INSTEAD, IT'S A SORT OF RECIPE DESCRIBING EXACTLY HOW TO CALCULATE A NUMBER FOR EACH VALUE OF L.

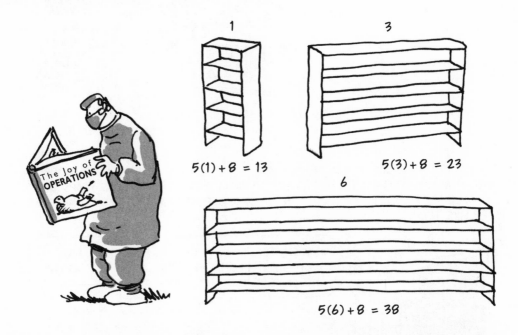

1
$5(1) + 8 = 13$

3
$5(3) + 8 = 23$

6
$5(6) + 8 = 38$

TO FIND THE VALUE OF $5L + 8$ FOR SOME VALUE OF L, YOU SUBSTITUTE (OR "PLUG IN") THAT NUMBER FOR L AND THEN DO THE ARITHMETIC AS SHOWN HERE. THIS IS CALLED **EVALUATING THE EXPRESSION** FOR A PARTICULAR **VALUE OF THE VARIABLE.**

Evaluation Example 1.

EVALUATE $p + .08p$ WHEN $p = 50$.

STEP 1. PLUG IN 50 WHEREVER YOU SEE p TO GET THE NUMERICAL EXPRESSION

$50 + (.08)(50)$

STEP 2. DO THE ARITHMETIC.

$50 + (.08)(50) = 50 + 4$

$= \mathbf{54}$

WE CAN ALSO EVALUATE EXPRESSIONS OF MORE THAN ONE VARIABLE, GIVEN VALUES FOR THOSE VARIABLES.

Evaluation Example 2. EVALUATE

$2(h + w)$ WHEN $h = 3$ AND $w = 7$.

STEP 1. PLUG IN THE VALUES TO GET

$2(3 + 7)$

STEP 2. DO THE ARITHMETIC.

$2(3 + 7) = 2 \times 10 = \mathbf{20}$

NO MEDICAL DEGREE REQUIRED!

43

SPEAKING "VARIABLESE"

LEARNING TO USE VARIABLES IS LIKE PICKING UP A NEW LANGUAGE. AT FIRST, EVERYTHING LOOKS STRANGE, BUT IN TIME THINGS BEGIN TO MAKE SENSE.

WHY LEARN THIS LANGUAGE? IN THE FIRST PLACE, VARIABLES ARE A HUGE HELP IN WRITING CLEAR MATHEMATICAL STATEMENTS. IN THE PRE-VARIABLE ERA (ROUGHLY THE YEARS BEFORE 1500), PEOPLE USED TO CALL AN UNKNOWN OR UNSPECIFIED QUANTITY THE "THING" AND SAY STUFF LIKE THIS:

ADD SIX TO *THING*, DOUBLE THE RESULT, AND SUBTRACT IT FROM FIVE TIMES *THING*, FORSOOTH, AND SITTEST THOU UP STRAIGHT.

TODAY WE WOULD WRITE A LETTER, SAY x, FOR "THING" AND EXPRESS THE WHOLE OPERATION LIKE SO:

$$5x - 2(x + 6)$$

WHICH I CAN GET EVEN KICKIN' BACK HERE IN THE LOUNGER...

44

IN THOSE EARLY DAYS, NOT EVERYONE LIKED HOW THE LITTLE LETTERS LOOKED.

"A scab of symbols as if a hen had been scratching there... they ought no more to appear in public, than the most deformed necessary business which you do in your chambers."

EWW!

BUT TO MOST MATHEMATICIANS, LETTER VARIABLES WERE A GIFT, A NEW TOY TOO PRECIOUS TO RESIST, AND HIGHLY USEFUL, TOO. "SYNCOPATED" (OR "SHORTENED") ALGEBRA OPENED UP ALL THE BEAUTIFUL MATH AND SCIENCE THAT FOLLOWED...

ANALYTIC GEOMETRY! CALCULUS! VECTOR SPACES! NUMBER THEORY! MEASURE THEORY! COMPLEX ANALYSIS! ALGEBRAIC TOPOLOGY! NETWORK THEORY! SYMBOLIC LOGIC! CELESTIAL MECHANICS! ELECTROMAGNETIC THEORY! SIGNAL ANALYSIS!

THIS MODERN MATH MADE THE MODERN WORLD. TRULY, WITHOUT ALGEBRA, WE WOULD HAVE NO ELECTRICITY, RADIO, TV, PHONES, MUSIC PLAYERS, COMPUTERS, AIRPLANES, MEDICAL IMAGING MACHINES, REFRIGERATORS, ROBOTS, ROCKETS...

A POX ON THE LOT OF IT!

LET'S DO A LITTLE WARM-UP EXERCISE WITH THIS NEW LANGUAGE BY DESCRIBING SOME **NEW SYMBOLS** IN TERMS OF VARIABLES. HERE THEY ARE NOW...

THANKS, GUYS!

THE SYMBOLS ARE RELATIVES OF THE FAMILIAR EQUALS SIGN =. < MEANS "IS LESS THAN," AND > MEANS "IS GREATER THAN."

SEE? THIS IS THE SMALL SIDE...

IN TERMS OF VARIABLES, WE'D PUT IT THIS WAY: SUPPOSE a AND b ARE **ANY TWO NUMBERS.**

$a < b$ MEANS THAT a IS TO THE LEFT OF b ON THE NUMBER LINE.

$a > b$ MEANS THAT a IS TO THE RIGHT OF b ON THE NUMBER LINE.

$a > 0$ SAYS THAT a IS POSITIVE, WHILE $a < 0$ SAYS THAT a IS NEGATIVE.

I < YOU!

AND I > YOU!

WE ALSO SOMETIMES USE THE SYMBOLS ≤, "IS LESS THAN OR EQUAL TO," AND ≥, "IS GREATER THAN OR EQUAL TO." SO

$a \geq 0$

MEANS THAT a COULD BE ANY POSITIVE NUMBER, OR POSSIBLY ZERO. a, WE WOULD SAY, IS **NON-NEGATIVE.** THE NON-POSITIVE NUMBERS WOULD BE THOSE NUMBERS b WITH $b \leq 0$.

$$a \geq 0$$

NON-NEGATIVE NUMBERS: ALL POSITIVE NUMBERS AND ZERO

NON-POSITIVE NUMBERS: ALL NEGATIVE NUMBERS AND ZERO

$$b \leq 0$$

WE CAN ALSO DESCRIBE A NUMBER'S **ABSOLUTE VALUE** MORE EASILY BY USING A VARIABLE. THE DEFINITION IS MUCH SHORTER THAN THE LONG-WINDED ONE GIVEN ON PAGE 18. IT'S THE SORT OF CLEVER DEFINITION A MATHEMATICIAN WOULD CALL "CUTE." IF a IS ANY NUMBER, ITS ABSOLUTE VALUE, $|a|$, IS DEFINED LIKE THIS:

CUTENESS IS IN THE EYE OF THE BEHOLDER...

$$|a| = a \text{ IF } a \geq 0$$
$$|a| = -a \text{ IF } a \leq 0$$

HOW CAN $|a|$ BE "NEGATIVE a" WHEN $|a|$ MUST BE POSITIVE? BECAUSE THE **NEGATIVE OF A NEGATIVE NUMBER IS POSITIVE!** (SEE PAGE 9.) IT MAY LOOK WEIRD, BUT IF a IS NEGATIVE $(a < 0)$, THEN $-a$ IS **POSITIVE**, AND $|a| = -a$. FOR INSTANCE, $|-5| = -(-5) = 5$.

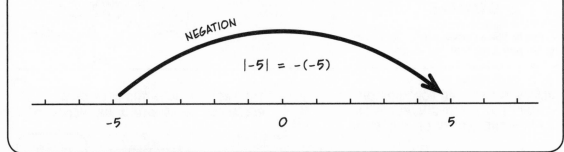

NEGATION

$$|-5| = -(-5)$$

-5   0   5

THESE SYMBOLS MAKE POSSIBLE A TOTALLY "ALGEBRAIC" DEFINITION OF THE ADDITION OF NEGATIVE NUMBERS IN TERMS OF THE FAMILIAR ADDITION AND SUBTRACTION OF POSITIVE NUMBERS THAT WE GREW UP WITH.

YEP! CHICKEN SCRATCHES...

IF $a > 0$ AND $b > 0$, THEN $a + b = |a| + |b|$

IF $a < 0$ AND $b < 0$, THEN $a + b = -(|a| + |b|)$

IF $a > 0$ AND $b < 0$, THEN

    IF $|a| > |b|$,
    THEN $a + b = |a| - |b|$

    IF $|a| < |b|$,
    THEN $a + b = -(|b| - |a|)$

47

LAWS OF COMBINATION

WHEN COMBINING NUMBERS OR VARIABLES, WE MUST ALWAYS FOLLOW THE LAW. OTHERWISE, WE MIGHT BE GUILTY OF GETTING THE WRONG ANSWER, AND THEN WHO KNOWS WHAT?

I HEREBY SENTENCE YOU TO 20 MINUTES OF CONFUSION AND ANXIETY.

THE FIRST LAWS SAY THAT IN **SOME** EXPRESSIONS THE ORDER OF **NUMBERS** DOESN'T MATTER:

COMMUTATIVE LAWS: IF *a* AND *b* ARE ANY TWO NUMBERS, THEN

$$a+b = b+a$$
$$ab = ba$$

(THIS IS REALLY TWO LAWS, ONE FOR SUMS AND ONE FOR PRODUCTS.)

WHEN ADDING OR MULTIPLYING **ONLY**, EITHER NUMBER CAN GO FIRST.

HERE'S THE PICTURE FOR ADDITION (DRAWN FOR POSITIVE NUMBERS ONLY). *a* + *b* IS THE LENGTH OF THIS STICK...

| a | b |

THEN I TURN IT AROUND...

TO MAKE *b* + *a*!

| b | a |

TURNING SOMETHING AROUND DOESN'T CHANGE ITS LENGTH, SO *a* + *b* = *b* + *a*.

THE PRODUCT *ab* IS THE AREA OF A RECTANGLE OF LENGTH *a* AND HEIGHT *b*.

AGAIN, TURN IT!

THE TIPPED-OVER RECTANGLE HAS AREA *ba*. TURNING SOMETHING DOESN'T CHANGE ITS AREA, SO *ba* = *ab*.

SOMETIMES THE ORDER OF **OPERATIONS** DOESN'T MATTER.

ASSOCIATIVE
LAWS: IF a, b, AND c ARE ANY THREE NUMBERS, THEN

$$(a+b)+c = a+(b+c)$$
$$(ab)c = a(bc)$$

WHEN ADDING OR MULTIPLYING **ONLY**, GROUPING ("ASSOCIATION") DOESN'T MATTER.

YOU DON'T NECESSARILY NEED FOUR HANDS FOR THIS, BUT IT HELPS!

Associative Examples:

1. $(2+3)+4 = 5+4 = 9$
$2+(3+4) = 2+7 = 9$

2. $(5 \times 3) \times 6 = 15 \times 6 = 90$
$5 \times (3 \times 6) = 5 \times 18 = 90$

THE PICTURE FOR ADDITION (OF POSITIVE NUMBERS) IS SUPER-SIMPLE. ALL THESE LINES OBVIOUSLY HAVE THE SAME TOTAL LENGTH. IT DOESN'T MATTER WHERE YOU BREAK THEM.

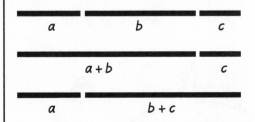

AND FOR MULTIPLICATION...

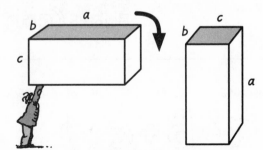

ONE BLOCK HAS VOLUME $(ab)c$. THE OTHER HAS VOLUME $a(bc)$. THESE MUST BE EQUAL, BECAUSE TURNING DOESN'T AFFECT VOLUME.

SO?

49

WHO NEEDS THESE SIMPLE-SOUNDING LAWS? ALL YOUR LIFE YOU'VE DONE SUMS WITHOUT THINKING ABOUT THE ORDER, WHEN THERE ARE ACTUALLY TWELVE DIFFERENT WAYS TO ADD THREE NUMBERS TOGETHER.

1. $a+(b+c)$
2. $(a+b)+c$
3. $a+(c+b)$
4. $(a+c)+b$
5. $b+(a+c)$
6. $(b+a)+c$
7. $b+(c+a)$
8. $(b+c)+a$
9. $c+(a+b)$
10. $(c+a)+b$
11. $c+(b+a)$
12. $(c+b)+a$

OUR TWO LAWS SAY THAT, GIVEN ANY CHOICE OF a, b, AND c, ALL THESE EXPRESSIONS HAVE THE SAME VALUE. FOR INSTANCE, TO SHOW THAT SUM #1 = SUM #7, WE REASON THIS WAY:

WHOA!

$$a+(b+c) = (b+c)+a$$

BY THE COMMUTATIVE LAW, SWITCHING THE NUMBERS a AND $b+c$

$$= b+(c+a)$$

BY THE ASSOCIATIVE LAW

BECAUSE THEY'RE ALL THE SAME, WE CAN REMOVE THE PARENTHESES AND SIMPLY WRITE

$$a+b+c$$

WITH NO RISK OF CONFUSION. THE SAME IS TRUE OF THE PRODUCTS $(ab)c$, $(ac)b$, ETC. JUST WRITE

$$abc$$

WITH NO PARENTHESES. AND YOU KNOW HOW I **LOVE** TO BURN PARENTHESES...

WE CAN ALSO SHUFFLE THE ORDER AND OMIT PARENTHESES IN SUMS OR PRODUCTS OF FOUR OR MORE NUMBERS. IT'S OKAY TO WRITE, FOR EXAMPLE,

$2abc$

WITHOUT CARING WHETHER IT MEANS $(2a)(bc)$, $2(a(bc))$, $((2a)b)c$, $(ab)(2c)$, OR ANY OF THE OTHER 116 (YES!) POSSIBILITIES. AND THE SAME WITH SUMS, OF COURSE.

HM... HOW MANY WAYS TO ADD SIX THINGS...?

THE POINT IS: IT DOESN'T MATTER!!

AT LEAST, NOT IN ALGEBRA 1...

THE PAYOFF IS THAT SUMS AND PRODUCTS OF NUMBERS AND VARIABLES BEHAVE EXACTLY AS YOU WOULD HOPE AND EXPECT. FOR INSTANCE, IF WE DOUBLE $3x$ WE HAD BETTER GET $6x$, AND THAT IS EXACTLY WHAT THE ASSOCIATIVE LAW GUARANTEES.

$$2(3x) = (2 \times 3)x$$
$$= 6x$$

AREN'T YOU RELIEVED?

FOR ADDITION, THE TWO LAWS GIVE THE SAME COMFORTING CONCLUSION: IF I ADD 3, SAY, TO $a + 2$, THEN I GET $a + 5$, JUST AS YOU'D THINK.

$$(a+2)+3 = a+(2+3)$$
$$= a+5$$

IN OTHER WORDS, THERE WILL BE NO AWFUL SURPRISES!

Minus Signs and the Laws

WE PICTURED THE COMMUTATIVE LAW $a+b=b+a$ WITH TWO POSITIVE NUMBERS, BUT THE LAW IS JUST AS TRUE WHEN a, b, OR BOTH ARE NEGATIVE. THAT'S BECAUSE ADDITION IS **DEFINED** TO BE THE SAME IN EITHER ORDER. (SEE P. 19 OR P. 47.) FOR INSTANCE,

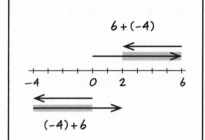

$$\left.\begin{array}{c} 6+(-4) \\ (-4)+6 \end{array}\right\}$$

SUBTRACT 4 FROM 6, THEN GIVE THE ANSWER THE SAME SIGN AS 6 BECAUSE $|6| > |-4|$.

SEEN WITH ARROWS, $6+(-4)$ TAKES 4 FROM THE **HEAD** END OF THE 6-ARROW, WHILE $-4+6$ TAKES 4 FROM THE **TAIL** END OF THE 6-ARROW. THE RESULT IS THE SAME: 2.

THE ASSOCIATIVE LAW ALSO APPLIES TO NEGATIVE NUMBERS.

THIS MAKES IT OKAY TO LEAVE OUT PARENTHESES IN "SUMS" THAT INCLUDE BOTH PLUS AND MINUS SIGNS, LIKE THIS ONE:

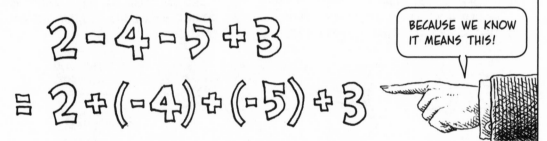

$$2-4-5+3$$
$$=2+(-4)+(-5)+3$$

BECAUSE WE KNOW IT MEANS THIS!

AND WE CAN WRITE THE EXPRESSION IN ANY ORDER, AS LONG AS THE MINUS SIGNS STICK TO THEIR NUMBERS. THESE ARE ALL THE SAME:

$$-4-5+3+2$$
$$-4+3-5+2$$
$$3+2-5-4$$
$$-5+2-4+3$$
$$2-4+3-5$$

ETC.!

HERE ARE TWO CONVENIENT WAYS TO EVALUATE OR SIMPLIFY A LONG SUM THAT INCLUDES SOME NEGATIVE NUMBERS.

1. GO LEFT TO RIGHT.

$$2-4-5+3$$
$$= -2-5+3$$
$$= -7 +3$$
$$= -4$$

2. GROUP NEGATIVES AND POSITIVES SEPARATELY AND ADD THE GROUPS. (ONLY ONE SUBTRACTION THIS WAY!)

$$2-4-5+3$$
$$= 2+3-4-5$$
$$= 5 - 9$$
$$= -4$$

WE CAN ALSO MOVE **VARIABLES** AROUND IN THE SAME WAY.

$$1+x-3 = x+1-3$$
$$= x-2$$

IN A **PRODUCT** OF SEVERAL NUMBERS AND/OR VARIABLES, WE CAN SHUFFLE AND REARRANGE TO BRING ALL MINUS SIGNS TO THE FRONT.

$$a(-2)(-3)(-b)$$
$$= a(-1)2(-1)3(-1)b$$
$$= (-1)(-1)(-1)(2)(3)ab$$
$$= (-1)6ab$$
$$= -6ab$$

BECAUSE $(-1)(-1) = 1$

BECAUSE $(-1)(-1) = 1$, WE GET THIS RULE: THE PRODUCT OF AN **EVEN** NUMBER OF MINUS SIGNS IS +; THE PRODUCT OF AN **ODD** NUMBER OF MINUS SIGNS IS −.

$$(-a)(-b)(-c)(-d)$$

FOUR MINUS SIGNS, EVEN $= abcd$

$$(-a)(-b)(c)(-d)$$

THREE MINUS SIGNS, ODD $= -abcd$

SO FAR, OUR LAWS HAVE ALLOWED SHUFFLING AND REGROUPING **WITHIN** SUMS AND PRODUCTS. OUR NEXT (AND FINAL) LAW OF COMBINATION IS DIFFERENT. IT DESCRIBES WHAT HAPPENS WHEN **TIMES** MEETS **PLUS**.

DISTRIBUTIVE LAW:

IF a, b AND c ARE ANY NUMBERS, THEN

$$a(b+c) = ab + ac$$

THE PRODUCT OF A NUMBER TIMES A SUM IS THE SUM OF THE TWO "PARTIAL PRODUCTS." MULTIPLICATION "DISTRIBUTES" OVER ADDITION (AGAIN, IT DOESN'T MATTER WHETHER THE NUMBERS a, b, AND c ARE POSITIVE, NEGATIVE, OR ZERO).

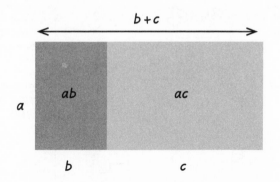

THE LARGE RECTANGLE, WITH AREA $a(b+c)$, IS MADE UP OF TWO SMALLER RECTANGLES WITH AREAS ab AND ac.

WE USE THE ASSOCIATIVE AND COMMUTATIVE LAWS ALMOST MINDLESSLY. OF **COURSE** $2+3 = 3+2$!! THE DISTRIBUTIVE LAW, ON THE OTHER HAND, CALLS FOR SOME CARE, BECAUSE WE'RE PUSHING A FACTOR ONTO MORE THAN ONE ITEM INSIDE THE PARENTHESES.

Numerical Example:

$$2(5 + 7) = 2(12) = 24$$

and also $= 2 \times 5 + 2 \times 7$

$$= 10 + 14$$

$$= 24$$

CHECK!

Examples with Variables:

1. $3(x + 1) = 3x + (3)(1) = 3x + 3$

2. $2a(x + 3) = 2ax + 6a$

NOTE THAT ORDER DOESN'T MATTER:

3. $P + \frac{1}{2}P = (1 + \frac{1}{2})P = \frac{3}{2}P$

4. $ax + 2x = (a + 2)x$

54

Multiplication distributes over long sums.

$$a(b+c+d+e+...) = ab+ac+ad+ae+...$$

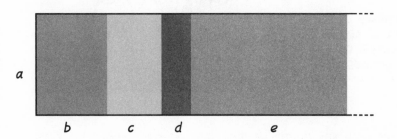

EEK!

Multiplication distributes over subtraction:

$$a(b-c) = ab-ac$$

THIS IS TRUE BECAUSE SUBTRACTION IS "NEGATIVE ADDITION."

$a(b - c) = a(b + (-c))$	DEFINITION OF SUBTRACTION
$= ab + a(-c)$	a DISTRIBUTES OVER ADDITION
$= ab + a((-1)c)$	$-c = (-1)c$
$= ab + (-1)(ac)$	SHUFFLING AND REGROUPING!
$= ab + (-ac)$	$(-1)ac = -ac$
$= ab - ac$	DEFINITION OF SUBTRACTION

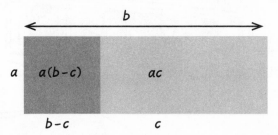

Negation distributes!

$$-(a+b) = -a-b$$

THIS IS TRUE BECAUSE NEGATION IS THE SAME AS MULTIPLICATION BY -1.

$$-(a + b) = (-1)(a + b)$$
$$= (-1)a + (-1)b$$
$$= -a - b$$

DISTRIBUTING A MINUS SIGN OVER A SUBTRACTION LOOKS LIKE THIS:

$$-(a-b) = -a+b$$
$$= b-a$$

NEGATING REVERSES ALL SIGNS!

MAGIC MINUS SIGN

SOMETIMES WE WANT TO "UNDISTRIBUTE," FOR EXAMPLE WHEN ADDING MULTIPLES OF A SINGLE VARIABLE SUCH AS $3x + 2x$.

3 APPLES PLUS 2 APPLES MAKE 5 APPLES, AND THE SAME FOR x!

$$3x + 2x = (3 + 2)x$$
$$= 5x$$

MULTIPLES OF A VARIABLE ADD AS EXPECTED. DIDN'T WE PROMISE NO AWFUL SURPRISES?

Reminder: THIS WORKS WITH ANY MULTIPLES, NOT JUST POSITIVE WHOLE NUMBERS. FOR INSTANCE,

$$2y - \frac{y}{2} = (2 - \frac{1}{2})y = \frac{3}{2}y \text{ OR } \frac{3y}{2}$$

2y

TAKE AWAY
y/2

LEAVES
$(1\frac{1}{2})y$ OR
3y/2

also:

$$P + .3P = (1 + .3)P = (1.3)P$$
(BECAUSE $P = 1 \cdot P$ WITH AN UNWRITTEN 1.)

$$6z - 2z = 4z$$

$$\frac{x}{2} + \frac{x}{3} = (\frac{1}{2} + \frac{1}{3})x = \frac{5}{6}x \text{ OR } \frac{5x}{6}$$

REAL-WORLD EXAMPLE: Discount Store

MY LOCAL BARGAIN STORE IS HAVING A 20%-OFF SALE. EVERYTHING IN THE STORE IS REDUCED BY AN AMOUNT EQUAL TO 20% OF THE MARKED PRICE.

I LOVE SALES, BUT WHAT DOES THIS HAVE TO DO WITH THE DISTRIBUTIVE LAW?

FINDING THE DISCOUNTED (SALE) PRICE OF AN ITEM WOULD SEEM TO TAKE TWO STEPS. STEP 1: FIND THE DISCOUNT BY MULTIPLYING THE PRICE BY .2 (THAT'S 20%, OR 20/100).

OKAY... $(.2) \times \$5$ EQUALS $\$1$...

STEP 2: SUBTRACT THE DISCOUNT FROM THE MARKED PRICE. IN TERMS OF A VARIABLE, AN ITEM WITH A MARKED PRICE P HAS A DISCOUNT OF $.2P$ AND A SALE PRICE OF

$$P - .2P$$

$\$5 - (.2) \times 5$
$= \$5 - \$1$
$= \$4.$
ALL RIGHT!

NOW APPLY THE DISTRIBUTIVE LAW. WE KNOW THAT $P = 1 \cdot P$ (THE 1 BEING UNWRITTEN), SO

$$P - .2P = (1 - .2)P = .8P$$

YES! $.8 \times \$5$ EQUALS $\$4$ TOO!

IN REALITY, THEN, WE CAN FIND THE DISCOUNTED PRICE IN JUST **ONE** STEP: **MULTIPLY THE MARKED PRICE BY .8!**

SO EASY!!!

EVEN BETTER: MAYBE WE WANT THE TOTAL COST OF SEVERAL ITEMS, SAY FOUR OF THEM WITH MARKED PRICES P, Q, R, AND S. THEIR SALES PRICES ADD UP TO $.8P + .8Q + .8R + .8S$, BUT—

$$.8P + .8Q + .8R + .8S$$
$$= .8(P + Q + R + S)$$

I MAY NOT AGREE WITH EVERY LAW, BUT THIS IS A GOOD ONE!

IN OTHER WORDS, TO FIND THE TOTAL COST OF SEVERAL ITEMS, ADD ALL THEIR MARKED PRICES AND MULTIPLY BY .8. YOU NEVER HAVE TO KNOW THE DISCOUNT OF A SINGLE ITEM!!

WHAT'LL YOU DO WITH ALL THE TIME YOU SAVE CALCULATING?

SHOP!

Problems

1. EVALUATE THESE NUMERICAL EXPRESSIONS:

a. $2 \times 3 + 1$

b. $2(3 + 1)$

c. $1 - \dfrac{4}{2} + 3(1 - \dfrac{1}{3})$

d. $5 - 3 + 2 - 4$

e. $5 - (3 + 2 - 4)$

f. $(1 - 2)/2$

g. $(2 - 100)/(40 + 9) - (-2)$

h. $\dfrac{9 - 4}{\frac{5}{3}}$

i. $(-6)(-5) - (-5)(6)$

j. $(\dfrac{1}{0.8})(40)$

k. $\dfrac{3.8 - 2(1 - 0.67)}{0.5}$

2. EVALUATE THE ALGEBRAIC EXPRESSIONS AT THE GIVEN VALUE(S) OF THE VARIABLE(S):

a. $5x - 4$ WHEN $x = 1$

b. $2P + 11$ WHEN $P = -6$

c. $\dfrac{3}{4}(3y - 1)(2y + 4)$ WHEN $y = 3$

d. $x + 2x + 3x - \dfrac{6}{x}$ WHEN $x = 1$.

3a. EVALUATE $2a(x + 1) - 3x + 4(a - 1)$ WHEN $x = 1$ AND $a = 2$.

b. EVALUATE THE SAME EXPRESSION WHEN $x = 2$ AND $a = 3$.

c. WHAT EXPRESSION IN a DO YOU GET IF YOU PLUG IN $x = 2$? CAN YOU SIMPLIFY IT BY USING THE DISTRIBUTIVE LAW?

4. SIMPLIFY THESE EXPRESSIONS USING THE DISTRIBUTIVE LAW. (THAT IS, DISTRIBUTE AND THEN COMBINE TERMS.)

a. $2(x + 5) - 1$

b. $3(x - 1) + 2(x + 1)$

c. $3(y + 2) + 4(y + 2)$

d. $3(2(2x - 1)) + 5) + x$

e. $1 - 2(1 - x)$

f. $a(1 - t) + 2a(2 - t)$

5. THE CHEAPO DEPOT CHANGED ITS DISCOUNT TO 15%. IF A PRICE TAG SAYS P DOLLARS, HOW MUCH IS THE SALE PRICE? YOU WANT A HAIRBRUSH ORIGINALLY PRICED AT $8.99 AND SOME GEL MARKED AT $4.95. WHAT WOULD THE TOTAL PRICE BE AFTER THE DISCOUNT?

6. USE THE ASSOCIATIVE LAW TO EXPLAIN WHY THE PRODUCTS IN EACH ROW ARE EQUAL. (HINT: DO YOU SEE ANY EVEN NUMBERS?)

$$2 \times 2 = 1 \times 4$$
$$4 \times 3 = 2 \times 6$$
$$6 \times 4 = 3 \times 8$$
$$8 \times 5 = 4 \times 10$$
$$10 \times 6 = 5 \times 12$$
$$12 \times 7 = 6 \times 14$$
$$14 \times 8 = 7 \times 16$$
$$\dots$$

7. A CREATIVE MATH TEACHER INVENTS A NEW OPERATION CALLED **RADDITION,** WRITTEN $a\#b$ (a RAD b) AND DEFINED BY $a\#b = a + b + ab$.

a. WHAT IS $4\#1$? $1\#4$?

b. IS RADDITION COMMUTATIVE? ASSOCIATIVE?

c. IF a IS ANY NUMBER, WHAT IS $a\#0$?

d. DOES MULTIPLICATION DISTRIBUTE OVER RADDITION? THAT IS, IS IT ALWAYS TRUE THAT $a(b\#c) = ab\#ac$?

8. IF R AND S ARE ROTATIONS OF A SPHERE (LIKE A BASKETBALL) AROUND ITS CENTER, MUST IT BE TRUE THAT $RS = SR$?

Chapter 5
Balancing Act

An algebraic expression is merely a recipe: it gives step-by-step instructions for operating on algebraic ingredients, in other words variables and numbers.

DOUBLE A NUMBER, THEN ADD 1, THEN TRIPLE THE RESULT.

$3(2x+1)$

An **EQUATION**, on the other hand, is a **STATEMENT**. It says that two different expressions **ARE THE SAME NUMBER**. Even though two expressions may not look alike, the equation says they result in the same value, once you do the arithmetic.

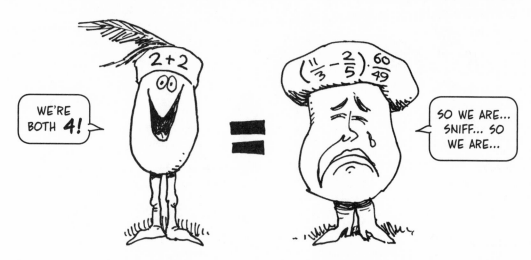

WE'RE BOTH **4**!

$2+2$

$=$

$\left(\frac{11}{3} - \frac{2}{5}\right) \cdot \frac{60}{49}$

SO WE ARE... SNIFF... SO WE ARE...

FOR INSTANCE, IN THE DISCOUNT STORE, THIS EXPRESSION DESCRIBES HOW TO CALCULATE THE SALE PRICE OF AN ITEM MARKED DOWN 20% FROM ITS ORIGINAL PRICE P.

MULTIPLY P TIMES .8.

$$.8P$$

WHEN THE CASHIER TELLS YOU WHAT YOU MUST ACTUALLY PAY, THAT'S AN EQUATION, A STATEMENT. IT SAYS THE SALE PRICE **IS EQUAL TO** SOME NUMBER.*

*NOT TAKING TAX INTO ACCOUNT. LET'S PRETEND THAT WE LIVE IN A MAGICAL TAX-FREE WORLD.

THAT'LL BE $5, PLEASE!

$$.8P = 5$$

OR MAYBE IT'LL BE $6!

LIKE ANY STATEMENT OF FACT, AN EQUATION CAN BE **TRUE** OR **FALSE**.

$$2 + 2 = 3 + 1 \quad \text{TRUE}$$

$$2 + 2 = 3 \quad \text{NOT SO TRUE!}$$

WE USE THE SYMBOL $\neq$ TO MEAN "IS **NOT** EQUAL TO," AS IN

$$2 + 2 \neq 3 \quad \text{TRUE}$$

AN EQUATION CONTAINING A **VARIABLE** MAY BE TRUE FOR SOME VALUE OR VALUES OF THE VARIABLE AND NOT FOR OTHERS. THE EQUATION $2x + 1 = 7$ IS TRUE WHEN $x = 3$ BECAUSE $2(3) + 1 = 7$, BUT FALSE WHEN $x = 4$ BECAUSE $2(4) + 1 = 9 \neq 7$.

A VALUE OF THE VARIABLE THAT MAKES THE EQUATION TRUE IS CALLED A

SOLUTION

OF THE EQUATION. A SOLUTION IS SAID TO

SATISFY

OR **SOLVE** THE EQUATION. $x = 3$ SATISFIES THE EQUATION $2x + 1 = 7$. TRY SOME OTHER VALUES OF x. DO YOU FIND ANY OTHER SOLUTIONS?

WHAT'S THE SOLUTION OF $.8P = 5$?

DUNNO. I JUST TOSSED THE TAG.

SUPPOSE YOU BOUGHT SOMETHING MARKED DOWN 20% TO $5. **WHAT WAS THE ITEM'S ORIGINAL PRICE BEFORE THE MARKDOWN?** UNFORTUNATELY, THE CASHIER THREW AWAY THE PRICE TAG AND LEFT YOU WITH NOTHING BUT AN EQUATION TO PONDER:

THE EQUATION TELLS YOU THE VALUE OF 80% OF P, A FRACTION OF P. HOW WOULD YOU FIND THE VALUE OF P ITSELF, ALL OF P, THAT IS $1 \cdot P$? HOW CAN YOU TURN $.8P$ INTO $1 \cdot P$? ANSWER: **MULTIPLY BY THE RECIPROCAL OF .8** (OR DIVIDE BY .8, SAME THING).*

THAT WILL CLEAR AWAY THE FACTOR .8, BECAUSE

$$\frac{1}{.8}(.8P) = (\frac{.8}{.8})P = P$$

BUT THEN WHAT ABOUT THE **5** ON THE OTHER SIDE OF EQUATION?

WELL, IF BEING A TRUE EQUATION MEANS ANYTHING, IT MEANS THIS: $.8P$ AND 5 ARE REALLY THE **SAME NUMBER.** NATURALLY, THEN, **ANY MULTIPLE** OF $.8P$ AND 5 WILL ALSO BE EQUAL TO EACH OTHER. HOW COULD THEY NOT BE? SO...

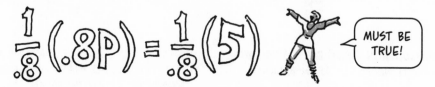

MUST BE TRUE!

NOW FOR THE ARITHMETIC:

$$\frac{1}{.8}(.8P) = (\frac{1}{.8})5$$

$$P = 5/.8 = \mathbf{6.25}$$

THE ORIGINAL PRICE WAS
$6.25.

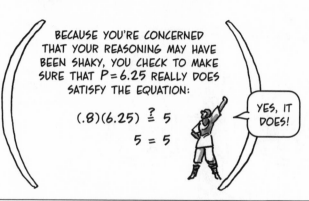

BECAUSE YOU'RE CONCERNED THAT YOUR REASONING MAY HAVE BEEN SHAKY, YOU CHECK TO MAKE SURE THAT $P = 6.25$ REALLY DOES SATISFY THE EQUATION:

$$(.8)(6.25) \stackrel{?}{=} 5$$

$$5 = 5$$

YES, IT DOES!

---

*IF YOU HATE THE DECIMAL, YOU CAN WRITE $.8 = 8/10 = 4/5$ AND ITS RECIPROCAL AS $5/4$.

WE HAVE JUST MET ALGEBRA'S FIRST **BIG IDEA**: GIVEN ANY TRUE EQUATION, YOU CAN "DO THE SAME THINGS TO BOTH SIDES" AND THE RESULTING EQUATION WILL STILL BE TRUE. THIS IDEA COMES FROM THE INVENTOR OF ALGEBRA HIMSELF, MUHAMMAD OF KHWARIZM, OR **AL-KHWARIZMI** (780–850).

YOU'RE VERY WELCOME!

AL-KHWARIZMI THOUGHT OF AN EQUATION AS "BALANCED." THE EXPRESSIONS ON THE TWO SIDES, THOUGH THEY LOOK DIFFERENT, EXPRESS THE SAME NUMBER.

IF WE **ADD** THE SAME THING (NUMBER, EXPRESSION, WHATEVER) TO BOTH SIDES, THE SIDES STILL BALANCE—THEY'RE STILL EQUAL TO ONE ANOTHER.

WE CAN ALSO **MULTIPLY** BOTH SIDES BY THE SAME THING AND STAY IN BALANCE.

WE CAN SOLVE MANY EQUATIONS USING ONLY THESE TWO STEPS, WHICH AL-KHWARIZMI CALLED "REBALANCING."

IT'S EASIER THAN IT LOOKS!

BEFORE GOING ON, LET ME SAY A FEW WORDS ABOUT THE LETTER MOST OFTEN USED AS A VARIABLE: x. THE POINT OF CHOOSING THIS MYSTERIOUS LETTER IS THAT IT STANDS FOR NOTHING IN PARTICULAR, WHETHER DISTANCE OR TIME OR PRICE. ALGEBRA, SAYS x, WORKS ON ANY VARIABLE, NO MATTER WHAT IT "MEANS." x CAN BE ANYTHING!

I'M A MASTER OF DISGUISE!

OKAY, TIME TO REBALANCE!

Example 1. SOLVE

$$4x + 5 = 2x + 11$$

$4x$ AND $2x$ ARE CALLED THE **VARIABLE TERMS**, WHILE THE "NAKED NUMBERS" 5 AND 11 ARE THE **CONSTANT TERMS**.

OH, DEAR! VARIABLES ON BOTH SIDES!

WHERE DO YOU START?

TO REBALANCE, WE CLEVERLY CHOOSE JUST THE RIGHT THINGS TO ADD OR SUBTRACT TO **REMOVE ALL VARIABLE TERMS** FROM THE **RIGHT** AND ALL **CONSTANT TERMS** FROM THE **LEFT**.

THIS HAS TO GO!

AND THIS!

SUBTRACTING 5 WILL CLEAR THE CONSTANT FROM THE LEFT, AND SUBTRACTING $2x$ WILL CLEAR THE VARIABLE TERM FROM THE RIGHT. LET'S DO IT—TO BOTH SIDES!

$$
\begin{array}{r}
4x + 5 = 2x + 11 \\
-5 \qquad -5 \\
-2x \qquad -2x \\
\hline
4x - 2x = 11 - 5 \\
2x = 6
\end{array}
$$

WE'RE ALMOST THERE! MULTIPLYING EVERYTHING BY 1/2, THE RECIPROCAL OF 2, WILL LEAVE x ALONE ON THE LEFT SIDE AND SOLVE THE EQUATION.

$$2x/2 = 6/2$$
$$x = 3$$

FINALLY, WE PLUG $x = 3$ INTO THE ORIGINAL EQUATION TO CHECK THAT IT REALLY IS A SOLUTION.

$$
\begin{array}{c}
4(3) + 5 \stackrel{?}{=} 2(3) + 11 \\
12 + 5 \stackrel{?}{=} 6 + 11 \\
17 = 17
\end{array}
$$

HOW TO SOLVE AN EQUATION, STEP BY STEP

(SOME EQUATIONS, ANYWAY)

1. "Prep" THE EQUATION IF NECESSARY BY GETTING RID OF PARENTHESES AND COMBINING LIKE TERMS. ("LIKE" MEANS THAT CONSTANTS ADD WITH CONSTANTS, VARIABLE TERMS WITH VARIABLE TERMS.)

> PARENTHESES ARE NOT OUR FRIENDS HERE!

2. Isolate, BY ADDITION AND/OR SUBTRACTION, THE CONSTANTS ON ONE SIDE (USUALLY THE RIGHT) AND THE VARIABLE TERMS ON THE OTHER (USUALLY THE LEFT).

> PUT 'EM IN THEIR PLACE!

3. Combine LIKE TERMS.

> SIMPLIFY! ALWAYS SIMPLIFY!

THE EQUATION WILL NOW LOOK LIKE THIS:
(SOME NUMBER)x = SOME OTHER NUMBER.

4. Multiply BOTH SIDES BY THE **RECIPROCAL** OF THE NUMBER IN FRONT OF THE VARIABLE. THIS NUMBER IS CALLED THE VARIABLE'S **COEFFICIENT.** FOR INSTANCE, GIVEN

$$4x = 12$$

4 IS THE COEFFICIENT OF x. MULTIPLYING BY $\frac{1}{4}$ GIVES

$$x = 3$$

AND THE EQUATION IS SOLVED.

> ISN'T THIS THE SAME AS DIVIDING BY THE COEFFICIENT?

> YEP!

5. Check THE ANSWER. THIS IS IMPORTANT, BOTH TO CHECK YOUR WORK AND FOR ANOTHER REASON TO BE EXPLAINED SHORTLY.

> IF IT CHECKS, THEN WE REALLY ARE DONE!

HERE'S A COMPLICATED EQUATION THAT NEEDS SOME PREP WORK TO SOLVE.

Example 2.

$$2(x-1)+3(x-2)+x = 2x+4$$

WE GO STEP BY STEP.

OH, YEAH! I'M REALLY DISGUISED IN THAT ONE!

1. THOSE PARENTHESES MAKE IT HARD TO SEE WHAT TO CLEAR FROM EACH SIDE, SO LET'S GET RID OF THEM. BY THE DISTRIBUTIVE LAW, $2(x-1) = 2x-2$ AND $3(x-2) = 3x-6$, MAKING THE EQUATION

$$2x-2 + 3x-6 + x = 2x+4$$

BIG DEAL! MY VALUE IS STILL SO OBSCURE...

COMBINING LIKE TERMS, VARIABLE WITH VARIABLE, CONSTANT WITH CONSTANT, GIVES:

$$6x-8 = 2x+4$$

PREP WORK DONE!

UH-OH... THEY'RE CLOSING IN...

2. NOW REBALANCING IS EASY: SUBTRACTING **2x** CLEARS THE VARIABLE TERM FROM THE RIGHT, AND ADDING **8** CLEARS THE CONSTANT FROM THE LEFT.

$$\begin{array}{rcl} 6x-8 &=& 2x+4 \\ -2x+8 && -2x+8 \\ \hline 6x-2x &=& 4+8 \end{array}$$

3. COMBINE TERMS: $6x-2x = 4x$ AND $4+8 = 12$. NOW THE EQUATION IS

$$4x = 12$$

4. DIVIDING BOTH SIDES BY 4, THE COEFFICIENT OF x, SOLVES IT.

$$x = 3$$

I ADMIT NOTHING UNTIL CHECKED...

5. FINALLY, CHECK THE ANSWER BY PLUGGING IN 3 FOR x THE ORIGINAL EQUATION:

$$2(3-1) + 3(3-2) + 3 \stackrel{?}{=} 2\cdot3 + 4$$
$$2\cdot2 + 3\cdot1 + 3 \stackrel{?}{=} 6 + 4$$
$$4 + 3 + 3 \stackrel{?}{=} 6 + 4$$
$$10 = 10$$

OH, ALL RIGHT... I **AM** 3... AND WAS ALL ALONG...

NEGATIVE COEFFICIENTS

AFTER REBALANCING, YOU MAY FIND A NEGATIVE COEFFICIENT STUCK TO THE UNKNOWN, LIKE THIS:

$$-3x = -9$$

AT THIS POINT, YOU COULD DIVIDE BY -3 (OKAY BUT MESSY), BUT IT'S JUST A LITTLE EASIER TO MULTIPLY BOTH SIDES BY -1 AND MAKE THE COEFFICIENT POSITIVE.

$$3x = 9$$

> JUST WAVE THE MAGIC MINUS WAND AND CHANGE THE SIGNS!

IN A SIMILAR WAY, ADDING **FRACTIONAL TERMS** CAN BE SLOW AND ANNOYING. LUCKILY, YOU CAN CLEAR ALL FRACTIONS FROM AN EQUATION BY MULTIPLYING THROUGH BY A COMMON DENOMINATOR. GIVEN AN EQUATION LIKE THIS ONE:

$$\frac{3}{2}x + \frac{1}{3} = \frac{5}{6}x + 2$$

MULTIPLY BOTH SIDES BY 6 (THE LEAST COMMON DENOMINATOR OF 2, 3, AND 6).

$$\frac{3 \times 6}{2}x + \frac{1 \times 6}{3} = \frac{5 \times 6}{6}x + (6)(2)$$

AFTER CANCELING COMMON FACTORS, ALL FRACTIONS ARE GONE!

$$9x + 2 = 5x + 12$$

THIS REBALANCES TO

$$4x = 10 \quad \text{AND SO}$$

$$x = \frac{5}{2}$$

TRY CHECKING THE SOLUTION.

> SIMPLIFY! ALWAYS SIMPLIFY!

A WORD ABOUT CHECKING: IT'S IMPORTANT FOR AN OBVIOUS REASON: PEOPLE MAKE MISTAKES!

I DO NOT MAKE MISTAKES. I MAKE EXCUSES.

THERE'S ANOTHER REASON AS WELL. IT HAS TO DO WITH ALGEBRA'S BASIC THINKING, WHICH ASSUMES, WHEN REBALANCING, THAT THE EQUATION WAS **TRUE.**

THE REASONING GOES LIKE THIS

THE EQUATION IS TRUE FOR SOME VALUE OF THE VARIABLE, THEN I CAN PUSH THINGS AROUND AND FIND OUT WHAT THAT VALUE MUST BE.

THAT'S A BIG "IF!"

HOW ABOUT THIS EQUATION?

$$x = x + 1$$

BLINDLY REBALANCING, WE CLEAR x FROM THE RIGHT

$$
\begin{array}{rr}
x = & x + 1 \\
-x & -x \\
\hline
0 = & 1
\end{array}
$$

AND CONCLUDE THAT $0 = 1$. OOPS!

THIS HAPPENED BECAUSE THE ORIGINAL EQUATION COULD NEVER HAVE BEEN TRUE IN THE FIRST PLACE. HOW CAN ANY NUMBER BE ONE MORE THAN ITSELF? THE EQUATION HAS **NO SOLUTION.**

CHECKING SOLUTIONS ASSURES US THAT THE ORIGINAL ASSUMPTION WAS OKAY: THE SOLVED EQUATION REALLY WAS TRUE FOR SOME VALUE OF THE VARIABLE.

QUICK REBALANCING, or CALL THE MOVERS!

NOW I'M GOING SHOW YOU A QUICKER WAY TO REBALANCE EQUATIONS. LET'S IMAGINE ANY EQUATION WHERE ONE SIDE IS THE SUM OF TWO EXPRESSIONS.

AND SUPPOSE WE WANT TO ELIMINATE FROM THE LEFT.

BY NOW, YOU KNOW THAT WE DO THIS BY SUBTRACTING THE EXPRESSION FROM BOTH SIDES. LET'S WRITE IT ON A LINE, RATHER THAN STACKED UP.

THE RESULT:

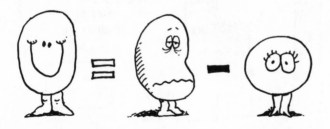

DO YOU SEE WHAT HAPPENS? THE EXPRESSION SEEMS TO JUMP FROM ONE SIDE TO THE OTHER, AND ITS SIGN CHANGES FROM PLUS TO MINUS!

 YOU CAN ALWAYS REBALANCE THIS WAY. INSTEAD OF WRITING OUT THE ADDITION OR SUBTRACTION OF TERMS, YOU CAN SIMPLY **MOVE** A TERM FROM ONE SIDE OF THE EQUATION TO THE OTHER AND FLIP ITS SIGN.

SO MOVING...

Example 3. SOLVE $x - 5 = 4x - 17$

AS ALWAYS, WE WANT TO ELIMINATE THE CONSTANT FROM THE LEFT AND THE VARIABLE TERM FROM THE RIGHT. WE COULD WRITE EVERYTHING OUT IN A STACK LIKE THIS...

$$
\begin{array}{rcl}
x - 5 &=& 4x - 17 \\
+ 5 && + 5 \\
-4x && -4x \\
\hline
\end{array}
$$

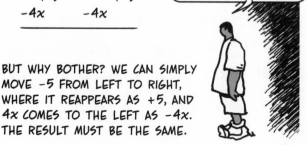

ALMOST LOOKS LIKE A GRADE-SCHOOL ARITHMETIC PROBLEM!

BUT WHY BOTHER? WE CAN SIMPLY MOVE -5 FROM LEFT TO RIGHT, WHERE IT REAPPEARS AS $+5$, AND $4x$ COMES TO THE LEFT AS $-4x$. THE RESULT MUST BE THE SAME.

$$x - 5 = 4x - 17$$

$$x - 4x = -17 + 5$$

$$-3x = -12$$

$$3x = 12$$

$$x = 4$$

AND WE CHECK:

$$4 - 5 \overset{?}{=} 4(4) - 17$$
$$-1 \overset{?}{=} 16 - 17$$
$$-1 = -1$$

NOW, YOUR MATH TEACHER MAY TELL YOU THAT MOVING TERMS AROUND LIKE THIS IS **OLD-FASHIONED!**

SINCE I WENT TO SCHOOL, SOME "EXPERT" DECIDED TO MAKE STUDENTS WRITE EVERYTHING OUT IN FULL, EVEN THOUGH MOVING TERMS IS QUICKER AND SHORTER...

FRANKLY, I THINK THE NEW WAY IS A BIG WASTE OF TIME! SINCE WHEN DO WE WANT TO GO **SLOWER** AND **WASTE PAPER?!!**

SO... WRITE IT OUT AND PLEASE YOUR TEACHER, OR SAVE A TREE AND DO IT MY WAY!

AND NOW, ON TO SOME PRACTICE PROBLEMS...

Problems

1. SOLVE THESE EQUATIONS (AND CHECK YOUR SOLUTIONS!):

a. $2x = x + 1$

b. $5x + 10 = 25$

c. $500x + 1,000 = 2,500$

(SUGGESTION: DIVIDE BOTH SIDES BY 500 BEFORE DOING ANYTHING ELSE.)

d. $7y - 1 = 5y + 9$

e. $3x + 4 = x - 5$

f. $4x + 1 = 7$

g. $4x + 1 = 0$

h. $1 - 2x = 3x - 19$

i. $2(1 - x) = 1 + x$

j. $2(60 - m) = 2(64 - 3m)$

k. $25 - 3x = 30 - 5x$

l. $\dfrac{t}{2} = \dfrac{t}{5} + \dfrac{3}{4}$

m. $\dfrac{P}{2} + \dfrac{P}{3} = 5$

n. $3(y - 1) + 2(y - 2) = y$

o. $6t = 4(t + 10)$

p. $\dfrac{x - 1}{2} + \dfrac{x - 2}{3} = \dfrac{1 + x}{6}$

2. SUPPOSE A PAIR OF SHOES IS MARKED DOWN BY 25%.

a. IF ITS ORIGINAL PRICE IS P, WRITE AN EXPRESSION FOR ITS SALE PRICE.

b. WRITE THE SAME EXPRESSION WITH A FRACTIONAL COEFFICIENT INSTEAD OF A DECIMAL.

c. IF THE SALE PRICE IS $66, WHAT WAS THE ORIGINAL PRICE?

d. IF THE SALE PRICE WAS Q, WRITE AN EXPRESSION FOR THE ORIGINAL PRICE IN TERMS OF THE VARIABLE Q.

3. SUPPOSE THE SALES TAX RATE IS 8% (THAT'S .08). THIS MEANS THAT THE TAX ON AN ITEM OF MARKED PRICE p IS $(.08)p$. THE TAX, OF COURSE, IS ADDED TO THE PRICE.

a. WHAT IS THE AFTER-TAX PRICE OF A CANDY BAR WITH A MARKED PRICE OF $1? $2? $p?

b. IF THE AFTER-TAX PRICE IS $3.78, WHAT WAS THE MARKED PRICE?

c. IF THE SALES TAX RATE IS r, WRITE AN EXPRESSION FOR THE AFTER-TAX PRICE OF AN ITEM PRICED AT p DOLLARS.

4. SUPPOSE a IS ANY NUMBER OTHER THAN 0. REBALANCE THIS EQUATION:

$$2ax + 3 = ax + 4$$

CAN YOU SOLVE FOR x? IN OTHER WORDS, CAN YOU WRITE AN EQUATION

$$x = \text{SOME EXPRESSION INVOLVING } a$$

THAT SATISFIES THE EQUATION?

5. FOLLOW THE 5-STEP PROGRAM TO SOLVE THIS EQUATION:

$$x + 1 = 1 + x$$

WHAT DID YOU "PROVE?" WHY DO YOU SUPPOSE THAT HAPPENED? DOES THIS EQUATION HAVE ANY SOLUTIONS? IF SO, WHAT ARE THEY?

Chapter 6
Real Wor(l)d Problems

To use algebra in daily life, we have to translate real situations into expressions and equations. In textbooks, these situations are called word problems, because they're described in words... but I prefer **WORLD** problems, because that's where they come from.

$2(l-x) = 3x + 7$

Example 1. KEVIN JUST FINISHED BUILDING A BOOKCASE. (HE'S STILL HAMMERING JUST BECAUSE IT FEELS SO GOOD.) IT'S 4 FEET TALL; IT HAS 5 SHELVES; AND IT USED A TOTAL OF 23 FEET OF BOARDS. HOW LONG IS EACH SHELF?

ASSUME THAT THE TOP SHELF IS BETWEEN THE UPRIGHTS, AS IN THE ILLUSTRATION, SO IT'S THE SAME LENGTH AS THE LOWER SHELVES.

BEGIN BY ORGANIZING THE INFORMATION INTO WHAT'S KNOWN AND WHAT'S UNKNOWN.

KNOWN:

SIDE HEIGHT: 4 FT.
NUMBER OF SIDES: 2
NUMBER OF SHELVES: 5
TOTAL LENGTH: 23 FT.

UNKNOWN:

SHELF LENGTH

WE KNOW SO MUCH!

SHELF LENGTH IS THE ONLY VARIABLE QUANTITY... SO CHOOSE AN ABBREVIATION THAT REMINDS YOU OF "LENGTH"...

LENGTH... AHEM... LLLENGGTH... HOW ABOUT "NGTH"?

MAY I MAKE A SUGGESTION?

LET'S USE L, SHALL WE?

NEXT WRITE AN ALGEBRAIC EXPRESSION FOR THE TOTAL LENGTH OF BOARDS IN TERMS OF L. WE DID THIS ON P. 39.

TOTAL LENGTH
=
$$5L + 8 \text{ FEET}$$

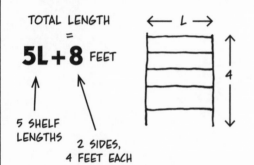

5 SHELF LENGTHS

2 SIDES, 4 FEET EACH

FINAL STEP OF THE SETUP: MAKE AN EQUATION. FOR THIS WE LOOK IN THE PROBLEM FOR A STATE-MENT, AND FIND THIS ONE: THE TOTAL LENGTH **IS EQUAL TO** 23 FEET.

$$5L + 8 = 23$$

THAT, RIGHT THERE, SAYS IT ALL!

WE SEEK THE VALUE (OR VALUES) OF L THAT MAKES THE EQUATION TRUE. IN OTHER WORDS, WE SEEK SOLUTIONS!

OBVIOUSLY, SOME VALUES OF L WILL MAKE 5L+8 DIF-FERENT FROM 23, BUT...

I GET IT! IT'S TIME TO SOLVE THE BLEEPIN' EQUATION!!

SO WE SOLVE!

$$5L + 8 = 23$$

$$5L = 23 - 8 \qquad \text{SUBTRACTING 8 FROM BOTH SIDES}$$

$$5L = 15 \qquad \text{ARITHMETIC}$$

$$L = \frac{15}{5} \qquad \text{DIVIDING BOTH SIDES BY 5}$$

$$L = 3 \qquad \text{ARITHMETIC}$$

R-RRIGHT... NOW WHAT WAS L AGAIN?

 RIGHT OVER THERE IT SAYS THAT L IS THE LENGTH OF ONE SHELF. SO WE'VE SHOWN THAT EACH SHELF IS 3 FEET LONG.

'COURSE, I KNEW THAT ALL ALONG, BUT THEN I BUILT THE THING.

THANKS FOR NOT SPOILING MY FUN BY TELLING ME...

AND WE CHECK:

$$5(3) + 8 \overset{?}{=} 23$$
$$15 + 8 \overset{?}{=} 23$$
$$23 = 23$$

Example 2. MOMO EARNS $2 MORE PER HOUR THAN CELIA. AFTER AN 8-HOUR SHIFT, THEIR COMBINED PAY IS $184. HOW MUCH DOES EACH OF THEM MAKE PER HOUR?

THE QUESTION IS—

WHAT ARE THE KNOWNS AND UNKNOWNS?

KNOWN:

TOTAL PAY, $184

TOTAL HOURS WORKED, 8

DIFFERENCE BETWEEN MOMO'S HOURLY AND CELIA'S HOURLY, $2

UNKNOWN:

CELIA'S HOURLY WAGE

MOMO'S HOURLY WAGE

NO, THE QUESTION IS—

DO WE NEED TWO DIFFERENT VARIABLES?

ALTHOUGH WE SEE TWO UNKNOWNS, WE DON'T HAVE TO ASSIGN LETTERS TO BOTH OF THEM, BECAUSE THEY'RE SO CLOSELY RELATED. LET'S START WITH CELIA'S HOURLY WAGE. CALL IT w. WE KNOW THAT MOMO'S HOURLY WAGE IS $2 MORE THAN CELIA'S, OR $w+2$.

w = CELIA'S HOURLY WAGE IN DOLLARS

$w+2$ = MOMO'S HOURLY WAGE IN DOLLARS

NO, THE QUESTION IS—

WHAT EXPRESSIONS DO WE WRITE?

THE PROBLEM TELLS US THE TOTAL PAY FOR 8 HOURS' WORK. LET'S WRITE EXPRESSIONS IN w FOR EACH GIRL'S EARNINGS IN 8 HOURS.

$8w$ CELIA'S EARNINGS

$8(w+2)$ MOMO'S EARNINGS

$8w + 8(w+2)$ TOTAL EARNINGS

THE EQUATION IS THE STATEMENT THAT THIS COMBINED AMOUNT EQUALS $184.

$$8w + 8(w+2) = 184$$

TO SOLVE IT, WE HAVE TO GET RID OF PARENTHESES.

$8w + 8w + 16 = 184$ — DISTRIBUTIVE LAW

$16w + 16 = 184$ — COMBINING TERMS

$16w = 168$ — SUBTRACTING 16 FROM BOTH SIDES

$w = \dfrac{168}{16}$ — DIVIDING BOTH SIDES BY 16

$w = 10.5$

AS BEFORE, WE HAVE TO REMEMBER WHAT w IS! THAT'S WHY WE **WROTE IT DOWN.** w = CELIA'S HOURLY WAGE. SO CELIA MAKES $10.50 PER HOUR, AND MOMO MAKES w+2 = $12.50 PER HOUR.

AND WE CHECK...

$8(10.5) + 8(12.5) \overset{?}{=} 184$

$84 + 100 \overset{?}{=} 184$

$184 = 184$ ✓

Example 3, Competing Claims.

CELIA AND JESSE DESIGN A FRIEND'S WEB SITE FOR A TOTAL PAY OF $180. CELIA THINKS SHE DID $120 WORTH OF WORK, AND JESSE THINKS HE DESERVES $80. UNFORTUNATELY, THEIR DEMANDS ADD UP TO $200...

HOW DO THEY DIVIDE UP THE MONEY?

LET'S TRY ALGEBRA FIRST, SHALL WE?

ONE WAY WOULD BE FOR EACH TO SACRIFICE THE SAME **AMOUNT**.

I'LL GIVE UP x DOLLARS IF YOU WILL! SOUND FAIR?

HM... NOT SURE...

LET'S SEE HOW THAT WOULD LOOK.

KNOWN:

CELIA WANTS 120

JESSE WANTS 80

TOTAL AVAILABLE IS 180

EACH GIVES UP THE SAME AMOUNT

UNKNOWN:

AMOUNT GIVEN UP

AMOUNT EACH GETS IN THE END

AGAIN WE START WITH A SINGLE VARIABLE, THE AMOUNT TO BE GIVEN UP. CALL IT x.

$$X = \text{AMOUNT TO BE CUT FROM EACH}$$

SIGH... I HATE CUTS, ESPECIALLY HAIRCUTS...

THESE EXPRESSIONS DESCRIBE HOW MUCH MONEY EACH PERSON WILL HAVE AFTER THE CUT.

CELIA WILL HAVE **120 − x**

JESSE WILL HAVE **80 − x**

THE EQUATION IS THE STATEMENT THAT THESE AMOUNTS MUST ADD TO $180.

$$(120 - x) + (80 - x) = 180$$

WHICH WE SOLVE EASILY.

$$200 - 2x = 180$$
$$2x = 200 - 180$$
$$2x = 20$$
$$x = 10$$

EACH SIDE WOULD GIVE UP $10. IN OTHER WORDS, THEY WOULD "SPLIT THE DIFFERENCE." (THE DIFFERENCE IS 20 AND THEY EACH GIVE UP HALF: 20/2 = 10.)

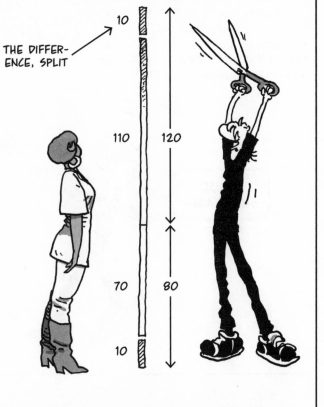

THE DIFFERENCE, SPLIT

KNOWING x, THE CUT, WE FIND EACH PERSON'S FINAL AMOUNT BY SUBTRACTING THE CUT FROM THE ORIGINAL CLAIM. CELIA WOULD GET $(120 - x) = $120 - $10 = $110, AND JESSE WOULD GET $(80 - x) = $80 - $10 = $70. FAIR? JESSE DOESN'T THINK SO!

WELL, **WHY NOT??!!** AFTER ALL THAT WORK!

HOW CAN I EXPLAIN THIS? 80/120 = 2/3... SO MY ORIGINAL CLAIM WAS 2/3 OF YOURS... BUT 70/110 IS A SMALLER FRACTION, OKAY???

AND IT'S TRUE.

$$\frac{70}{110} < \frac{80}{120}$$

AFTER SPLITTING THE DIFFERENCE, JESSE WOULD GET LESS **RELATIVE TO CELIA.**

ONE WAY TO SETTLE THE PROBLEM OF COMPETING CLAIMS IS TO SPLIT THE DIFFERENCE, WHICH WE PICTURED AS A PROCESS OF CUTTING. EACH CLAIM WAS CUT BY THE SAME AMOUNT. NOW JESSE SUGGESTS ANOTHER POSSIBILITY.

EQUAL SHRINKAGE!

HERE'S THE IDEA: THE TWO CLAIMS ADD UP TO $200, LIKE THIS.

120   80

PLEASE RECALL: ON PAGE 28, WE SAID THAT SCALING UP AND DOWN IS DONE BY **MULTIPLICATION.**

NOW INSTEAD OF CUTTING OFF BITS, IMAGINE **SQUEEZING** THAT PICTURE...

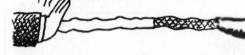

UNTIL ITS LENGTH SHRINKS TO 180, THE AMOUNT OF MONEY ACTUALLY AVAILABLE.

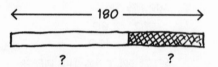
←——— 180 ———→

?   ?

NOW THE SPLIT LOOKS THE SAME AS BEFORE, ONLY SMALLER, LIKE A REDUCED PHOTO.

IN OTHER WORDS, WE WANT TO MULTIPLY BOTH CLAIMS BY THE SAME **SHRINKAGE FACTOR,** A NUMBER THAT IS STILL UNKNOWN. CALL IT r, FOR RATE.

r = SHRINKAGE FACTOR

WE MULTIPLY THIS FACTOR TIMES EACH CLAIM TO GET THE FINAL SETTLEMENT AMOUNT.

CELIA'S AMOUNT: **$120r$**

JESSE'S AMOUNT: **$80r$**

AS BEFORE, THE EQUATION SAYS THAT THE SUM OF THESE IS EQUAL TO $180.

$$120r + 80r = 180$$

THIS IS AN EASY EQUATION TO SOLVE.

$$120r + 80r = 180$$
$$200r = 180$$
$$r = \frac{180}{200}$$
$$r = \frac{9}{10}$$

WELL, WE DIDN'T SHRINK TOO MUCH...

NOW CELIA GETS

$$120r = \frac{9}{10}(120) = \mathbf{\$108}$$

AND JESSE GETS

$$80r = \frac{9}{10}(80) = \mathbf{\$72}$$

WHICH ADD TO $180, WHICH CHECKS THE SOLUTION.

NOTE THAT JESSE DID BETTER THIS WAY, AND CELIA DID WORSE, THAN BY SPLITTING THE DIFFERENCE.

WELL, OKAY! $2 MORE FOR THE BIG GUY!

BUT DOESN'T THAT MEAN I GAVE UP MORE OF MY CLAIM THAN HE DID?

CELIA IS RIGHT. THIS WAY, SHE SEES $12 CUT FROM HER ORIGINAL CLAIM, AND ONLY $8 CUT FROM JESSE'S.

COMPETING CLAIMS CAN ALSO ARISE WHEN SOMEBODY DIES IN DEBT. BIG BOB HERE WAS HAVING HIS HOUSE REMODELED WHEN HE HAD THE BAD LUCK TO EXPIRE, OWING FRED THE BUILDER $2.5 MILLION ($2,500,000). MEANWHILE, BIG BOB'S HOUSEKEEPER RITA SAYS SHE WAS PROMISED HALF A MILLION ($500,000) ON ACCOUNT OF THEIR "VERY SPECIAL RELATIONSHIP." UNFORTUNATELY, THERE'S ONLY $1 MILLION IN BOB'S BANK ACCOUNT. HOW DO THEY SETTLE?

AND WHY ARE YOU GOING THROUGH BOB'S POCKETS?

THE TWO CLAIMS TOTAL $3 MILLION. IF FRED AND RITA SPLIT THE DIFFERENCE, THEN EACH OF THEM WOULD GIVE UP HALF THE DIFFERENCE BETWEEN THE TOTAL CLAIM AND THE AMOUNT AVAILABLE. CALL THIS NUMBER x.

$$x = \tfrac{1}{2}(\$3,000,000 - \$1,000,000)$$
$$= \$1,000,000$$

BLINDLY FOLLOWING THE FORMULA, FRED THE BUILDER CALCULATES HIS SHARE AS

$2.5 MILLION $- x =$
$2.5 MILLION $-$ $1 MILLION =
$1.5 MILLION

AND RITA THE HOUSEKEEPER "GETS"

$500,000 $- x =$
$500,000 $-$ $1,000,000 =
$-$500,000

YEP, THAT'S A **NEGATIVE** HALF MILLION!

SPLITTING THE DIFFERENCE LITERALLY WOULD FORCE RITA TO **PAY** $500,000, WHICH FRED WOULD POCKET IN ADDITION TO THE MILLION DOLLARS FROM DEAD BOB! FAIR?

DEATH IS MORE CRUEL THAN I KNEW.

IN REAL LIFE, OF COURSE, THIS WOULD NEVER HAPPEN. AT WORST, RITA WOULD GET NOTHING, AND FRED THE BUILDER WOULD TAKE THE WHOLE MILLION, WELL SHORT OF WHAT HE'S OWED.

#%$*& %$(#!*&% $#$&...

#%$*&% %$(#!*&%...

ON THE OTHER HAND, THEY COULD DIVIDE THE ESTATE BY APPLYING A SHRINK FACTOR r TO THEIR CLAIMS. THIS WAY FRED WOULD GET $2,500,000 r, RITA WOULD GET $500,000 r, AND THESE NUMBERS MUST TOTAL $1 MILLION.

$$2,500,000r + 500,000r = 1,000,000$$

$$5r + r = 2 \qquad \text{DIVIDING BOTH SIDES BY } 500,000$$

$$6r = 2$$

$$r = \frac{1}{3}$$

TO THE NEAREST DOLLAR, THEN, FRED'S TAKE WOULD BE

$$\frac{1}{3}(\$2,500,000) \approx \$833,333$$

AND RITA'S:

$$\frac{1}{3}(\$500,000) \approx \$166,667$$

RITA DOES GET SOMETHING, BUT FRED LOSES EVEN MORE MONEY. THE BUILDER IS NOW OUT $2,500,000 - $833,333 = $1,666,667.

AND SHE LOSES NOTHING BUT A PROMISE!

THIS PROBLEM ALSO COMES UP IN BANKRUPTCY, WHEN A PERSON OR COMPANY GOES BROKE OWING MONEY... AND I HOPE YOU SEE THAT MATH ALONE CAN'T DECIDE WHAT'S "FAIR" EVERY TIME. THAT'S WHY BANKRUPTCY AND INHERITANCE ARE HANDLED BY LAW COURTS, NOT BY MATH PROFESSORS.

EVERYONE EQUALLY UNHAPPY? THEN JUSTICE IS DONE!

Problems

1. MOMO OWES $5 TO JESSE AND $10 TO KEVIN, BUT SHE HAS ONLY $9 IN HER POCKET. HOW MUCH SHOULD SHE PAY SO THAT EACH RECEIVES THE SAME FRACTION OF WHAT HE'S OWED?

2. CELIA EARNS $2/HR MORE THAN MOMO. MOMO MAKES AS MUCH IN 10 HOURS AS CELIA MAKES IN 8 HOURS. HOW MUCH DO THEY EACH MAKE PER HOUR?

3. JESSE EARNS $3/HR MORE THAN KEVIN. AFTER WORKING 8 HOURS, JESSE GIVES KEVIN 10% OF HIS PAY, AFTER WHICH THEY HAVE EQUAL AMOUNTS. WHAT ARE THEIR HOURLY WAGES?

4. A PICTURE FRAME IS TWICE AS TALL AS IT IS WIDE. THE TOTAL LENGTH OF THE WOOD THAT WENT INTO IT WAS 66 INCHES. HOW LONG ARE THE SIDES OF THE FRAME?

5. A PICTURE FRAME IS 4/3 AS WIDE AS IT IS LONG. THE TOTAL LENGTH OF LUMBER WAS 303 INCHES, BUT THERE WAS A 9" PIECE LEFT OVER. WHAT ARE THE DIMENSIONS OF THE FRAME?

6. IF A DISCOUNTED ITEM ORIGINALLY PRICED AT $A HAS A SALE PRICE OF $B, WHAT IS THE PERCENT DISCOUNT, IN TERMS OF A AND B?

CAN YOU BREAK A C?

7a. WRITE AN EXPRESSION FOR THE AMOUNT OF MONEY n NICKELS ARE WORTH.

7b. WRITE AN EXPRESSION FOR THE AMOUNT OF MONEY m DIMES ARE WORTH.

7c. IF I HAVE TWICE AS MANY DIMES AS NICKELS, AND THE AMOUNT OF MONEY IS $1.75, HOW MANY NICKELS DO I HAVE? HOW MANY DIMES?

8. JESSE STARTS WITH $4.00. HE GIVES CELIA SOME QUARTERS AND HALF AS MANY DIMES, ENDING UP WITH $1.60. HOW MANY QUARTERS AND DIMES DID HE GIVE?

9. A ROW OF TREES EXTENDS AWAY FROM A HOUSE. THE DISTANCE FROM TREE #1 TO TREE #2 IS TWICE THE DISTANCE FROM THE HOUSE TO TREE #1; THE DISTANCE FROM TREE #2 TO TREE #3 IS TWICE THE DISTANCE FROM TREE #1 TO TREE #2; AND SO IT GOES, THE DISTANCE BETWEEN EACH ADJACENT PAIR OF TREES BEING TWICE THE DISTANCE BETWEEN THE PREVIOUS PAIR. IF THE DISTANCE FROM THE HOUSE TO THE FIFTH TREE IS 930 FEET, HOW FAR FROM THE HOUSE IS THE FIRST TREE?

10. BIG AL AND LITTLE BENNY ROB A BANK. AL GIVES BENNY $1,000 AND KEEPS $2,738 FOR HIMSELF. WHEN BENNY COMPLAINS, AL MAKES HIM AN OFFER: THE NEXT MONEY THEY STEAL WILL BE SPLIT 1/4 TO AL AND 3/4 TO BENNY, UNTIL BENNY HAS HALF AS MUCH AS AL. HOW MUCH DO THEY NEED TO STEAL BEFORE BENNY'S TOTAL REACHES HALF AL'S?

Chapter 7
More Than One Unknown

REALITY IS FULL OF VARIABLES. HEIGHTS AND WEIGHTS RISE AND FALL... PRICES GO UP AND (SOMETIMES) DOWN... THE WORLD IS ALWAYS CHANGING IN COUNTLESS WAYS... SO SHOULDN'T WE LET AT LEAST ONE MORE VARIABLE INTO OUR EQUATIONS AND GET A LITTLE MORE **REAL**?

$$3 = 5x - 7 + y$$

LET'S START WITH A **CARPENTRY PROJECT**. CELIA GOES TO THE HARDWARE STORE FOR SOME NAILS. SHE NEEDS TWO DIFFERENT KINDS, BRASS AND IRON.

Hammer & Nail
SALON

Brass
3¢ EA.

Iron
2¢ EA.

FOR SOME REASON OR OTHER, SHE TOSSES THEM ALL INTO THE SAME BAG...

AND TAKES THEM TO KEVIN'S WOOD SHOP.

KEVIN IS NOT HAPPY! HE WANTS TO KNOW HOW MANY NAILS **OF EACH KIND** CELIA BOUGHT!

AND DON'T MAKE ME COUNT 'EM ALL, PLEASE!

KEVIN'S FIRST IDEA IS TO **WEIGH** THE NAILS. THE SCALE TELLS HIM THEY WEIGH 900 GRAMS. HE ALSO FINDS THAT ONE BRASS NAIL WEIGHS 3 GRAMS, WHILE ONE IRON NAIL WEIGHS 4 GRAMS.

NOW HE GETS ALGEBRAIC. HE LETS

B = THE NUMBER OF BRASS NAILS

I = THE NUMBER OF IRON NAILS

THEN $3B$ IS THE WEIGHT OF ALL THE BRASS NAILS, IN GRAMS, AND $4I$ IS THE WEIGHT OF ALL THE IRON NAILS, ALSO IN GRAMS. THE SUM OF THESE EXPRESSIONS IS THE TOTAL WEIGHT, 900 GRAMS, AND THIS STATEMENT BECOMES AN EQUATION.

(1) $$3B + 4I = 900$$

KEVIN TRIES SOLVING FOR B.

$$3B + 4I = 900 \quad \text{(EQUATION 1)}$$

$$3B = 900 - 4I \quad \text{(SUBTRACTING } 4I \text{ FROM BOTH SIDES)}$$

$$(2) \quad B = 300 - \frac{4}{3}I \quad \text{(DIVIDING BOTH SIDES BY 3)}$$

INSTEAD OF FINDING A NUMBER, HE GETS AN EXPRESSION INVOLVING I. IN EQUATION (2), KEVIN HAS SOLVED FOR B "IN TERMS OF I." SO WHAT'S B? KEVIN AND CELIA ARE LEFT SCRATCHING THEIR HEADS.

IN FACT, **MANY** POSSIBLE VALUES OF B AND I SOLVE EQUATION 2. FOR INSTANCE, IF WE SIMPLY **GUESS** THAT $I = 30$, WE CAN PLUG THAT INTO (2) AND FIND

$$B = 300 - \frac{4}{3}(30)$$

$$= 300 - 40$$

$$= 260$$

AND THIS PAIR OF VALUES, $I = 30$, $B = 260$, SOLVES EQUATION 1, AS WE SEE BY PLUGGING THEM IN.

$$3(260) + 4(30)$$

$$= 780 + 120$$

$$= 900$$

IF WE HAD PICKED SOME OTHER VALUE OF I, SAY $I = 93$, THEN

$$B = 300 - \frac{4}{3}(93)$$

$$B = 300 - 124 = 176$$

AND YOU CAN CHECK THAT THIS PAIR OF VALUES ALSO SOLVES EQUATION (1).

FOR ANY VALUE OF I, THERE IS A CORRESPONDING VALUE OF B. THE EQUATION HAS **MANY SOLUTIONS.**

HERE ARE A FEW SOLUTIONS—BUT BY NO MEANS ALL OF THEM!

I	B	$3B + 4I$
3	296	900
6	292	900
9	288	900
12	284	900
93	176	900
99	168	900
...	...	
200	$33\frac{1}{3}$	900

YOU CAN BUY $\frac{1}{3}$ OF A NAIL?

WELL, YOU CAN **IMAGINE** IT...

CAN KEVIN EVER FIND B AND I USING ALGEBRA? JUST MAYBE... BECAUSE CELIA HAS SOME **MORE INFORMATION:** SHE REMEMBERS THE **COST** OF THE NAILS!

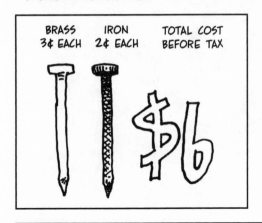

BRASS 3¢ EACH	IRON 2¢ EACH	TOTAL COST BEFORE TAX

NOW YOU TELL ME...

KEVIN SETS UP A NEW EQUATION. ALL NUMBERS ARE IN CENTS.

$3B$ = COST OF BRASS NAILS
$2I$ = COST OF IRON NAILS

THE TOTAL COST = 600 CENTS OR $6.00, IS THE SUM OF THESE.

(3) $3B + 2I = 600$

STILL USING A PENCIL INSTEAD OF A HAMMER!

THE FIRST EQUATION WITH TWO VARIABLES HAS MANY SOLUTIONS. CAN IT BE THAT A SECOND EQUATION WILL NARROW THE POSSIBILITIES TO A SINGLE PAIR OF NUMBERS, THE ACTUAL ANSWER? CAN WE FIND VALUES OF B AND I THAT SATISFY **BOTH** EQUATIONS **AT THE SAME TIME?**

IF THE ANSWER IS "NO," YOU ARE REALLY WASTING MY TIME...

I WOULD NEVER!

TWO EQUATIONS IN TWO VARIABLES

START WITH TWO
EQUATIONS LIKE THIS:

$$ax + by = e$$
$$cx + dy = f$$

IF I RUN OUT OF
LETTERS, I'LL USE
PIECES OF FRUIT!

HERE a, b, c, d, e, AND f CAN BE ANY NUMBERS, AND x AND y ARE THE VARIABLES. NOTE THAT THERE ARE NO TERMS INVOLVING PRODUCTS OF VARIABLES LIKE xy OR xx OR x/y, JUST x AND y ALONE, WITH CONSTANT COEFFICIENTS. WE'VE JUST SEEN AN EXAMPLE: CELIA AND KEVIN'S CARPENTRY EQUATIONS (NOW USING x AND y INSTEAD OF B AND I).

$$3x + 4y = 900$$
$$3x + 2y = 600$$

WE WILL NOW SHOW THREE DIFFERENT WAYS TO SOLVE THIS PAIR OF EQUATIONS—THREE! AND THESE THREE WAYS ARE CALLED...

SUBSTITUTION, ELIMINATION,

AND, UM, WELL, THE THIRD WAY DOESN'T EXACTLY HAVE A NAME...

Substitution

AGAIN, START WITH THESE EQUATIONS:

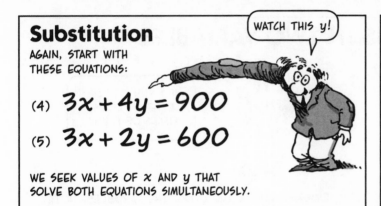

WATCH THIS y!

(4) $3x + 4y = 900$

(5) $3x + 2y = 600$

WE SEEK VALUES OF x AND y THAT SOLVE BOTH EQUATIONS SIMULTANEOUSLY.

START BY USING EQUATION 5 TO SOLVE FOR y IN TERMS OF x.

(5) $\quad 3x + 2y = 600$

$\qquad 2y = 600 - 3x$

(6) $\quad y = 300 - \dfrac{3}{2}x$

SINCE y IS THE SAME AS THAT EXPRESSION IN x, WE CAN **SUBSTITUTE** IT FOR y IN EQUATION 4.

'SCUSE ME!

SAME AS y!

NOW WE HAVE AN EQUATION IN ONE VARIABLE, x ALONE.

$$3x + 4(300 - \tfrac{3}{2}x) = 900$$
$$3x + 1200 - 6x = 900$$
$$6x - 3x = 1200 - 900$$
$$3x = 300$$
$$x = 100$$

AND y? THE VERY FIRST STEP WAS TO FIND y IN TERMS OF x.

(6) $\quad y = 300 - \dfrac{3}{2}x$

$\qquad y = 300 - \dfrac{3}{2}(100)$

$\qquad y = 300 - 150$

$\qquad y = 150$

SO THE ANSWER IS

$x = 100$ (BRASS NAILS)
$y = 150$ (IRON NAILS)

GREAT! TIME TO POUND!

AND WE CHECK THAT THIS SOLUTION SOLVES **BOTH** EQUATIONS.

(4) $3(100) + 4(150) \overset{?}{=} 900$
$\qquad 300 + 600 = 900$

(5) $3(100) + 2(150) \overset{?}{=} 600$
$\qquad 300 + 300 = 600$

AS PROMISED!

HITS THE NAIL ON THE HEAD!

Elimination

THE SUBSTITUTION METHOD GETS RID OF ("ELIMINATES") ONE VARIABLE IN A RATHER ROUNDABOUT WAY. THIS METHOD GOES STRAIGHT TO IT!

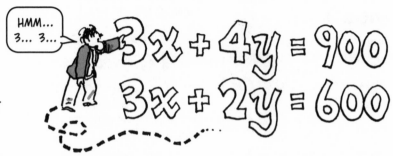

IF WE SUBTRACT THE LEFT SIDE OF ONE EQUATION FROM THE LEFT SIDE OF THE OTHER, THOSE $3x$ TERMS WILL CANCEL OUT. LET'S TRY THAT SUBTRACTION.

$$3x + 4y = 900$$
$$- (3x + 2y = 600)$$
$$\overline{}$$

$4y - 2y \Rightarrow \quad 2y = 300 \quad \leftarrow 900 - 600$

SUBTRACTING LEFT FROM LEFT AND RIGHT FROM RIGHT, WE'RE STILL TAKING AWAY EQUAL THINGS FROM EQUAL THINGS—SO THE RESULTS MUST BE EQUAL TOO!

NOW x IS ELIMINATED, AND WE HAVE FOUND

$$2y = 300$$
$$y = 150$$

FIND x BY PLUGGING IN 150 FOR y IN EITHER OF THE ORIGINAL EQUATIONS.

(4) $3x + 4y = 900$

   $3x + 4(150) = 900$

   $3x + 600 = 900$

   $3x = 300$

   $x = 100$

Method 3

THIS ONE MIGHT BE CALLED "FIND y TWICE." THE EQUATIONS

(4) $3x + 4y = 900$

(5) $3x + 2y = 600$

ALLOW US TO SOLVE FOR y IN TERMS OF x IN TWO DIFFERENT WAYS.

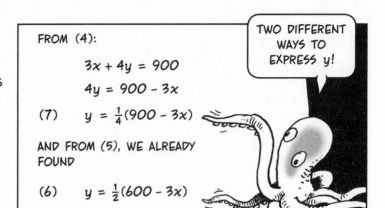

FROM (4):

$$3x + 4y = 900$$

$$4y = 900 - 3x$$

(7) $y = \frac{1}{4}(900 - 3x)$

AND FROM (5), WE ALREADY FOUND

(6) $y = \frac{1}{2}(600 - 3x)$

TWO DIFFERENT WAYS TO EXPRESS y!

THE EXPRESSIONS $\frac{1}{4}(900-3x)$ AND $\frac{1}{2}(600-3x)$ ARE BOTH EQUAL TO y, SO THEY MUST BE EQUAL TO EACH OTHER.

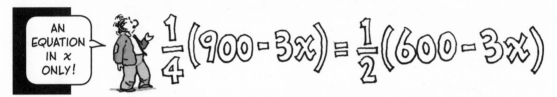

AN EQUATION IN x ONLY!

$$\frac{1}{4}(900-3x) = \frac{1}{2}(600-3x)$$

WE SOLVE IT WITHOUT MUCH FUSS:

$$\frac{1}{4}(900 - 3x) = \frac{1}{2}(600 - 3x)$$

$$900 - 3x = 2(600 - 3x)$$

(MULTIPLYING BY 4 TO CLEAR FRACTIONS)

$$900 - 3x = 1200 - 6x$$

$$6x - 3x = 1200 - 900$$

$$3x = 300$$

$$x = 100$$

FIND y FROM EITHER (6) OR (7) BY PLUGGING IN 100 FOR x.

$$y = \frac{1}{2}(600 - 3x)$$

$$y = \frac{1}{2}(600 - (3)(100))$$

$$y = \frac{1}{2}(300)$$

$$y = 150$$

COULDN'T WE GET A DIFFERENT ANSWER ONCE IN A WHILE?

More About Elimination

IN THE BRASS-TACKS PROBLEM, BOTH EQUATIONS HAD A TERM, $3x$, THAT WAS EASY TO ELIMINATE BECAUSE IT HAD THE SAME COEFFICIENT, 3, IN BOTH EQUATIONS. IT'S ALSO EASY TO ELIMINATE A VARIABLE WHEN ITS COEFFICIENTS ARE **DIFFERENT**, LIKE, SAY, THESE:

(8) $$5x + 2y = 13$$

(9) $$2x + 3y = 14$$

> SUBTRACTING WON'T ELIMINATE x OR y!

> YET!

THE IDEA IS TO MULTIPLY THE EQUATIONS BY NUMBERS THAT PRODUCE THE SAME COEFFICIENT FOR ONE OF THE VARIABLES. HERE, FOR INSTANCE, WE CAN MULTIPLY THE TOP EQUATION BY 3 AND THE BOTTOM BY 2, PRODUCING THE TERM $6y$ IN BOTH.

$$3 \times (5x + 2y = 13)$$
$$2 \times (2x + 3y = 14)$$

$$\Longrightarrow$$

$$15x + 6y = 39$$
$$4x + 6y = 28$$

NOW SUBTRACT AS BEFORE, AND THE $6y$ TERMS WILL CANCEL.

$$
\begin{array}{r}
15x + 6y = 39 \\
-(\ 4x + 6y = 28\ ) \\
\hline
11x \qquad = 11 \\
x \qquad = 1
\end{array}
$$

THEN FIND y BY SUBSTITUTING $x = 1$ IN EITHER EQUATION 8 OR 9:

(9) $2x + 3y = 14$

$2(1) + 3y = 14$

$3y = 12$

$y = 4$

YOU CAN CHECK THE SOLUTION!

> THE NEXT BEST SOLUTION TO A SOAP SOLUTION!

> I PERSONALLY PREFER ELIMINATION TO THE OTHER TWO TECHNIQUES. IT'S NEATER AND LESS ERROR-PRONE... AND AS YOU CAN SEE ON P. 89, DESCRIBING IT IN CARTOONS MAKES FOR A MUCH CLEANER PAGE LAYOUT, TOO, ALWAYS A GOOD SIGN...

> CLEAN PAGE, CLEAN MIND, CLEAN MATH!

ELIMINATION WORKS JUST AS WELL WITH NEGATIVE COEFFICIENTS. FOR EXAMPLE,

$$8x - 5y = 1$$
$$3x + 2y = 12$$

MULTIPLY BY 2 →
MULTIPLY BY 5

$$16x - 10y = 2$$
$$15x + 10y = 60$$

$$31x \qquad = 62$$
$$x \qquad = 2$$

THEN **ADD** (NOT SUBTRACT) TO ELIMINATE y!

IT'S LEFT TO YOU TO FIND y AND CHECK THE ANSWER.

CAUTION: SOMETIMES TWO EQUATIONS CAN LEAD TO FRUSTRATION. FOR INSTANCE, USING ELIMINATION ON

$$x + y = 2$$
$$2x + 2y = 4$$

GIVES

$$0 = 0$$

NOT TOO HELPFUL! THAT'S BECAUSE THE SECOND EQUATION IS SIMPLY TWICE THE FIRST ONE. EVERY SOLUTION OF THE FIRST (AND THERE ARE MANY) ALSO SOLVES THE SECOND.

ON THE OTHER HAND, THE PAIR

$$x + y = 3$$
$$x + y = 2$$

LEADS BY SUBTRACTION TO

$$0 = 1$$

A SURE SIGN SOMETHING IS WRONG! HERE THE EQUATIONS HAVE NO SOLUTION IN COMMON. HOW CAN TWO NUMBERS x AND y ADD TO 2 AND ALSO ADD TO 3? NO WAY...

IN THE NEXT CHAPTER, THIS WILL BECOME CLEARER, AS WE DRAW PICTURES OF EQUATIONS...

MORE?

THE SAME TECHNIQUES CAN BE USED TO SOLVE THREE EQUATIONS IN THREE UNKNOWNS.

$$(10) \quad x + y + 2z = 4$$

$$(11) \quad 2x + y + z = 3$$

$$(12) \quad 3x + 4y + 2z = 10$$

WE CAN ELIMINATE y, FOR INSTANCE, FROM THE PAIR (10) AND (11)...

$$
\begin{array}{ll}
(10) & x + y + 2z = 4 \\
(11) & -(\,2x + y + z = 3\,) \\
\hline
(13) & -x \quad\;\; + z = 1
\end{array}
$$

AND ALSO ELIMINATE y FROM THE PAIR (10) AND (12).

$$
\begin{array}{ll}
(4 \times \text{EQ'N } 10) & 4x + 4y + 8z = 16 \\
(12) & -(\,3x + 4y + 2z = 10\,) \\
\hline
(14) & x \quad\;\; + 6z = 6
\end{array}
$$

(13) AND (14) ARE A PAIR OF EQUATIONS IN TWO VARIABLES (x AND z), WHICH WE CAN SOLVE AS BEFORE.

$$
\begin{array}{ll}
(13) & -x + z = 1 \\
(14) & x + 6z = 6 \\
\hline
& 7z = 7 \\
& z = 1
\end{array}
$$

PLUG z = 1 INTO (13) TO FIND x.

$$
\begin{array}{ll}
(13) & -x + 1 = 1 \\
& x = 0
\end{array}
$$

SOLVE FOR y, THE ONLY VARIABLE LEFT, BY PLUGGING THESE VALUES OF x AND z INTO ANY ONE OF THE ORIGINAL EQUATIONS.

$$
\begin{array}{ll}
(10) & 0 + y + (2)(1) = 4 \\
& y = 4 - 2 \\
& y = 2
\end{array}
$$

THESE VALUES SOLVE ALL THREE EQUATIONS. (YOU SHOULD CHECK!)

AND SO ON FOR 4 EQUATIONS IN 4 VARIABLES, 5 EQUATIONS IN 5 VARIABLES, 6 EQUATIONS IN...

WHERE DOES IT ALL END?

THIS CHAPTER? RIGHT HERE!

Problems

SOLVE THESE SETS OF EQUATIONS.

1. $x + y = 51$
   $x - y = 3$

2. $r + s = 104$
   $r - s = 5$

3. $6x + 9y = 42$
   $15x - 2y = 7$

4. $2p + 4q = -18$
   $3p - 4q = 3$

5. $\dfrac{x}{2} + 4y = \dfrac{5}{2}$
   $x + 7y = 1$

6. $6.9r - 4.2s = 14.7$
   $2r + 2.4s = 18.5$

7. $2p + 4q = -18$
   $3p - 4q = 3$

8. $\dfrac{1}{3}x - \dfrac{1}{2}y = 5$
   $\dfrac{1}{2}y - \dfrac{1}{4}x = 7$

9. $2t + 3u + 2v = -1$
   $-6t - 5u - v = -11$
   $10t + u - v = 31$

10. $2x + 3y + 10z = 16$
    $3x + 2z = 10$
    $5x - 3y = 2$

11a. FIND TWO NUMBERS WHOSE SUM IS 23 AND WHOSE DIFFERENCE IS 5.

  b. FIND TWO NUMBERS WHOSE SUM IS 1,026 AND WHOSE DIFFERENCE IS 18.

12. A FISHING BOAT BRINGS IN TWO KINDS OF FISH, BASS AND COD. THE PRICE AT THE DOCK IS $2.25 PER POUND FOR BASS AND $1.85 PER POUND FOR COD. TODAY'S CATCH TOTALED 5,000 POUNDS AND SOLD FOR $10,450. HOW MANY POUNDS OF EACH KIND OF FISH WERE CAUGHT?

13. FIND TWO NUMBERS WHOSE SUM IS 12.476 AND WHOSE DIFFERENCE IS 17.511.

14. IF I DOUBLE JESSE'S AGE AND ADD IT TO CELIA'S, IT COMES TO 44. IF I DOUBLE CELIA'S AGE AND ADD IT TO JESSE'S, THE SUM IS 43. HOW OLD ARE CELIA AND JESSE?

15. MOMO HAS $7.00 WORTH OF NICKELS AND QUARTERS. THERE ARE 64 COINS ALTOGETHER. HOW MANY OF EACH COIN DOES SHE HAVE?

16. A TRUCK, STARTING WITH A FULL TANK OF FUEL, DRIVES A LOAD OF SAND TO A CONSTRUCTION SITE. ALONG THE WAY, SAND SLOWLY LEAKS OUT A HOLE IN THE FLOOR. WHEN THE TRUCK ARRIVES, ITS WEIGHT IS FOUND TO BE 110 POUNDS SHORT.

WORKERS FILL THE TANK WITH FUEL AND BILL THE DRIVER $24.80. IF THEY CHARGE $4.00 PER GALLON OF FUEL AND $.06 PER POUND OF LOST SAND, HOW MUCH SAND LEAKED OUT OF THE TRUCK? HOW MUCH FUEL DID IT BURN? ASSUME A GALLON OF FUEL WEIGHS 6 POUNDS.

17. SOLVE FOR x AND y IN TERMS OF a.

   $ax + 2y = 3$
   $x + y = 2$

Chapter 8
Drawing Equations

IN CASE YOU WERE WONDERING, THIS ISN'T THE FIRST BOOK TO MAKE CARTOONS ABOUT ALGEBRA... NO... THAT HONOR GOES ALL THE WAY BACK TO THE EARLY 1600s, WHEN FRENCHMAN

RENÉ DESCARTES

("REH-**NAY** DAY-**CART**") FIRST TURNED ALGEBRA INTO DRAWINGS THAT WE MIGHT CALL "DESCARTOONS."

THIS ONE ALWAYS CRACKS ME UP!

DESCARTES WANTED TO DRAW THE RELATIONSHIP BETWEEN TWO VARIABLES, SO INSTEAD OF ONE NUMBER LINE, HE DREW TWO... AND RATHER THAN RUNNING THEM SIDE BY SIDE, HE CROSSED THEM AT THEIR ZERO POINTS.

NOW THE WHOLE PLANE BECOMES A GRID. EVERY POINT IS PROVIDED WITH AN "ADDRESS" CONSISTING OF TWO NUMBERS IN ORDER, LIKE THIS: (x, y). THE FIRST NUMBER SAYS WHERE THE POINT IS HORIZONTALLY ON THE GRID, THE SECOND NUMBER WHERE IT IS VERTICALLY. THE POINT WHERE THE NUMBER LINES CROSS, CALLED THE **ORIGIN,** HAS ADDRESS $(0, 0)$.

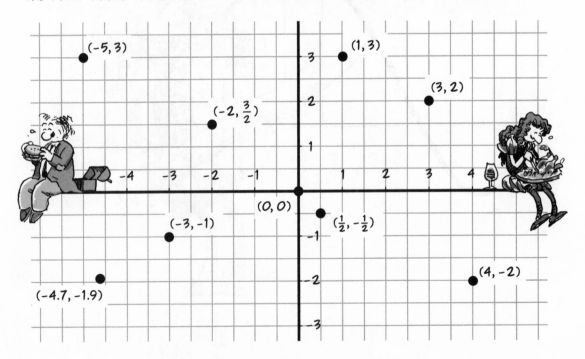

THE HORIZONTAL NUMBER LINE IS OFTEN CALLED THE **x-AXIS** AND THE VERTICAL NUMBER LINE THE **y-AXIS**. THE TWO NUMBERS OF A POINT'S ADDRESS ARE CALLED ITS **x-COORDINATE** AND ITS **y-COORDINATE**. TO FIND A POINT'S x-COORDINATE, FOLLOW A VERTICAL LINE FROM THE POINT TO THE x-AXIS; TO FIND ITS y-COORDINATE, GO HORIZONTALLY FROM THE POINT TO THE y-AXIS.

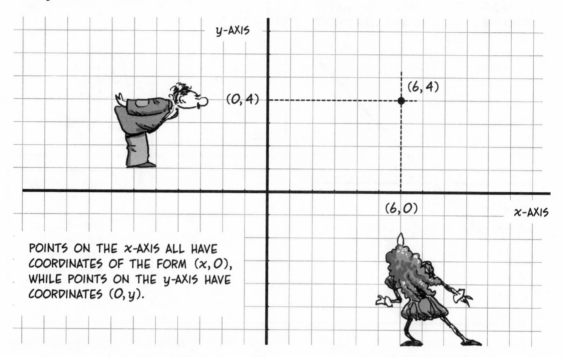

POINTS ON THE x-AXIS ALL HAVE COORDINATES OF THE FORM $(x, 0)$, WHILE POINTS ON THE y-AXIS HAVE COORDINATES $(0, y)$.

IF A CITY WERE LAID OUT LIKE THIS (AND MANY ARE—CHECK OUT A MAP OF NEW YORK CITY'S MANHATTAN), YOU MIGHT SAY THAT THE POINT (x, y) IS AT THE INTERSECTION OF x AVENUE AND y STREET. OF COURSE, OUR "CITY" HAS FRACTIONAL AND IRRATIONAL STREETS, TOO...

97

NOW LET'S PICTURE A SIMPLE EQUATION, $y = x$. A PAIR (x, y) SATISFIES THIS EQUATION WHEN THE TWO COORDINATES x AND y ARE **EQUAL** TO EACH OTHER. WE MARK ALL THE POINTS WHERE $x = y$, SUCH AS $(0, 0)$, $(1, 1)$, $(-3.14, -3.14)$, AND ALL THE REST. THESE POINTS LIE ON A STRAIGHT LINE CALLED THE EQUATION'S **GRAPH**.

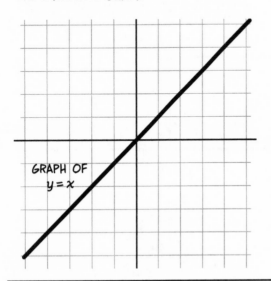

GRAPH OF
$y = x$

NEXT WE "GRAPH" THE EQUATION $y = 2x$. BEGIN BY FILLING IN A SMALL TABLE WITH A FEW VALUES OF x AND y. WE CAN PICK ANY OLD VALUES OF x; IT DOESN'T MATTER.

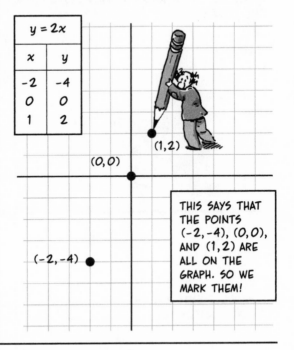

$y = 2x$	
x	y
-2	-4
0	0
1	2

$(1, 2)$

$(0, 0)$

$(-2, -4)$ ●

THIS SAYS THAT THE POINTS $(-2, -4)$, $(0, 0)$, AND $(1, 2)$ ARE ALL ON THE GRAPH. SO WE MARK THEM!

IF YOU LAY A STRAIGHTEDGE AGAINST THOSE POINTS, YOU'LL FIND THAT THEY LIE ON A STRAIGHT LINE. **EVERY POINT ON THE LINE** SATISFIES THE EQUATION $y = 2x$. THIS LINE IS THE GRAPH OF $y = 2x$. AS YOU SEE, IT'S STEEPER THAN THE GRAPH OF $y = x$.

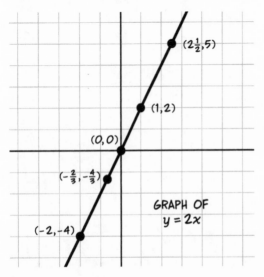

$(2\frac{1}{2}, 5)$

$(1, 2)$

$(0, 0)$

$(-\frac{2}{3}, -\frac{4}{3})$

GRAPH OF
$y = 2x$

$(-2, -4)$

HERE ARE THE GRAPHS OF EQUATIONS $y = mx$ FOR SEVERAL VALUES OF m. ALL OF THEM PASS THROUGH THE ORIGIN (WHY?), AND LARGER VALUES OF m MAKE STEEPER LINES. WHEN m IS NEGATIVE, THE GRAPH SLOPES "BACKWARD," I.E., IT GOES DOWN TO THE RIGHT.

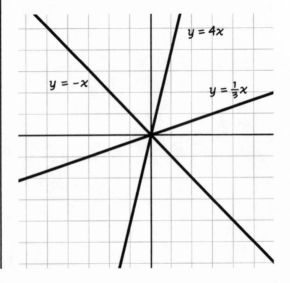

$y = 4x$

$y = -x$

$y = \frac{1}{3}x$

HOW ABOUT THE EQUATION $y = x + 2$? GIVEN ANY VALUE OF x, ADD 2 TO FIND y. STARTING AT ANY POINT x ON THE x-AXIS, GO x UNITS VERTICALLY, THEN UP TWO UNITS.

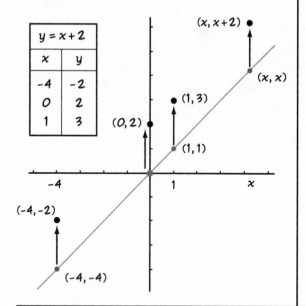

$y = x + 2$	
x	y
-4	-2
0	2
1	3

I HOPE YOU CAN SEE THAT THIS GRAPH LOOKS EXACTLY LIKE THE GRAPH OF THE EQUATION $y = x$, BUT SHIFTED TWO UNITS UPWARD.

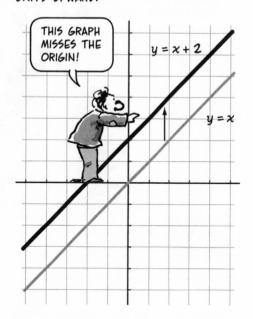

THIS GRAPH MISSES THE ORIGIN!

GIVEN ANY NUMBER a, THE GRAPH OF $y = x + a$ LOOKS LIKE THE GRAPH OF $y = x$ SHIFTED VERTICALLY a UNITS (UP IF $a > 0$, DOWN IF $a < 0$).

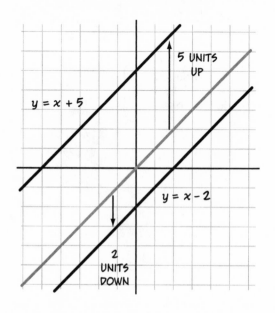

IN THE SAME WAY, IF m IS ANY NUMBER, THE GRAPH OF

$$y = mx + b$$

IS IDENTICAL TO THE GRAPH OF $y = mx$, BUT SHIFTED b UNITS VERTICALLY.

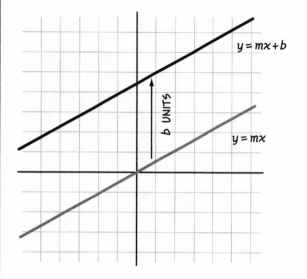

NOW LET'S THINK A LITTLE BIT ABOUT STEEPNESS OR SLOPE. MAYBE YOU'VE SEEN A ROAD SIGN LIKE THIS ONE: A **10% UPHILL GRADE** MEANS THE ROAD CLIMBS 0.1 MILE FOR EACH MILE OF FORWARD (HORIZONTAL) PROGRESS.

WFF!

10%
next 5 mi

0.1 MILE

1 MILE

THE GREATER THE HEIGHT YOU HAVE TO CLIMB (THE "RISE") FOR A GIVEN AMOUNT OF FORWARD PROGRESS (THE "RUN"), THE GREATER YOUR EFFORT AND THE STEEPER THE SLOPE.

A LINE'S SLOPE, THEN, IS MEASURED BY DIVIDING ITS RISE BY ITS RUN.

$$\text{SLOPE} = \frac{\text{rise}}{\text{run}}$$

I'M RISING, BUT I'M SURE AS HECK NOT RUNNING...

OH, MAN...

RISE

RUN

IN THE USA, INTERSTATE HIGHWAY GRADES AREN'T SUPPOSED TO EXCEED 6%, BUT LOCAL ROADS MAY BE STEEPER... AND OUR THOUGHT-LINES CAN BE STEEPER STILL. A GRADE OF **100%** WOULD RISE ONE MILE EVERY MILE (OR A FOOT EVERY FOOT, OR TEN METERS EVERY TEN METERS, ETC).

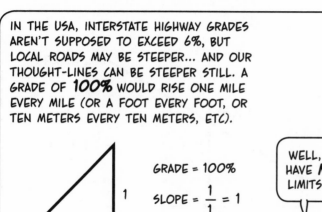

GRADE = 100%

$$\text{SLOPE} = \frac{1}{1} = 1$$

WELL, I HAVE **MY** LIMITS...

AND WE CAN GO EVEN STEEPER... THERE'S NO LIMIT ON THE SLOPE!

SLOPE CAN GO DOWNHILL, TOO, AND THE IDEA IS THE SAME—THE FARTHER YOU DROP FOR EACH UNIT OF FORWARD PROGRESS, THE STEEPER IT IS, IN A **NEGATIVE SENSE.**

THIS TAKES THE OPPOSITE OF EFFORT!

WHETHER UPHILL OR DOWN, THE MATH IS THE SAME: DIVIDE THE CHANGE IN ALTITUDE (POSITIVE OR NEGATIVE) BY THE FORWARD PROGRESS. GOING DOWNHILL, THE "RISE" IS REALLY A DROP, SO IT COUNTS AS NEGATIVE.

"RISE"

RISE IS NEGATIVE
RUN IS POSITIVE

$$\text{SLOPE} = \frac{\text{RISE}}{\text{RUN}} \text{ IS NEGATIVE}$$

RUN

NOW LET'S SEE HOW THIS LOOKS USING ALGEBRA.

Slope and Intercept

GIVEN AN EQUATION $y = mx + b$, WHAT DO THE NUMBERS m AND b TELL US ABOUT THE GRAPH? START WITH b.

WHEN $x = 0$, THE EQUATION SAYS

$$y = m(0) + b$$
$$= b$$

SO THE POINT $(0, b)$ IS ON THE LINE. IN OTHER WORDS, b IS WHERE **THE LINE CROSSES THE y-AXIS.** THE NUMBER b IS CALLED THE **y-INTERCEPT** OF THE LINE.

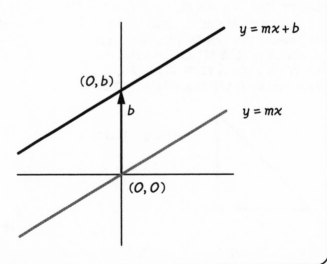

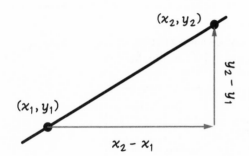

AND NOW m: WE SAW ON P. 99 THAT m HAS SOMETHING TO DO WITH A LINE'S SLOPE, SO LET'S CALCULATE THE SLOPE. WE TAKE TWO POINTS ON THE LINE—ANY TWO POINTS—AND DIVIDE THE RISE BY THE RUN. IF THE POINTS HAVE COORDINATES (x_1, y_1) AND (x_2, y_2),* THEN THE RISE IS $y_2 - y_1$ AND THE RUN IS $x_2 - x_1$.

LYING ON THE LINE, BOTH POINTS' COORDINATES SATISFY THE EQUATION.

$$y_1 = mx_1 + b$$
$$y_2 = mx_2 + b$$

SUBTRACT y_1 FROM y_2 TO FIND THE RISE.

$$y_2 - y_1 = mx_2 - mx_1 \quad \text{(b CANCELS)}$$
$$= m(x_2 - x_1) \quad \text{(DISTRIBUTIVE LAW)}$$

IN WORDS: RISE OVER RUN EQUALS m!

SINCE $x_2 - x_1$ ISN'T ZERO, WE CAN DIVIDE BOTH SIDES BY THIS NUMBER TO GET THE RISE OVER THE RUN:

$$\frac{y_2 - y_1}{x_2 - x_1} = m$$

THAT MONSTER ON THE LEFT IS CALLED THE **DIFFERENCE QUOTIENT.** THIS EQUATION SAYS THAT THE DIFFERENCE QUOTIENT IS **ALWAYS m,** FOR **ANY** TWO POINTS ON THE LINE. **m IS THE SLOPE!!**

*READ AS "EX-ONE," "WYE-ONE," "EX-TWO," "WYE-TWO." NO ARITHMETIC IS IMPLIED BY THE LITTLE SUBSCRIPT NUMBERS $_1$ AND $_2$. THEY ARE SIMPLY LABELS TO IDENTIFY WHICH OF THE TWO DIFFERENT POINTS "OWN" THE COORDINATES. x_1 IS THE x-COORDINATE OF THE FIRST POINT, AND SO ON.

Example. HERE IS THE GRAPH OF $y = 2x - 1$ WITH SOME OF ITS POINTS' COORDINATES LISTED IN A TABLE. NOTE THAT EVERY TIME x INCREASES BY 1, y INCREASES BY 2, THE COEFFICIENT OF x.

IN FACT, **ANY** TWO POINTS ON THIS LINE HAVE A DIFFERENCE QUOTIENT OF **2**. LET'S TRY $(-2, -5)$ AND $(2, 3)$.

$$\frac{y_2 - y_1}{x_2 - x_1} = \frac{3 - (-5)}{2 - (-2)}$$

$$= \frac{8}{4} = \mathbf{2}$$

TRY THIS WITH OTHER PAIRS OF POINTS. IF YOU GET ANYTHING OTHER THAN 2, YOU MADE A MISTAKE!

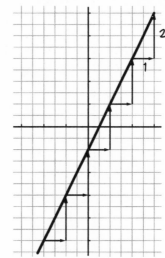

x	$2x-1$
-3	-7
-2	-5
-1	-3
0	-1
1	1
2	3
3	5

THE EQUATION $y = mx + b$ IS IN

SLOPE-INTERCEPT form.

WE CAN READ A GRAPH'S SLOPE AND y-INTERCEPT DIRECTLY FROM THIS EQUATION. IT TELLS US HOW STEEP THE GRAPH IS, AND HOW FAR ABOVE OR BELOW THE ORIGIN IT PASSES. IT ALSO DIRECTLY GIVES US THE VALUE OF y FOR EACH VALUE OF x.

SUCH A BEAUTIFUL, BEAUTIFUL FORM...

WHY, THANKS!

Example. GRAPH THE EQUATION

$$6x - 2y = 5$$

AND FIND ITS SLOPE AND y-INTERCEPT.

SOLUTION: FIRST USE ALGEBRA TO REWRITE THE EQUATION IN SLOPE-INTERCEPT FORM.

$$6x - 2y = 5$$

$$-2y = -6x + 5$$

$$y = 3x - \frac{5}{2}$$

THE SLOPE IS 3; THE y-INTERCEPT IS $-5/2$; YOU MAKE A TABLE; I'LL DRAW THE GRAPH!

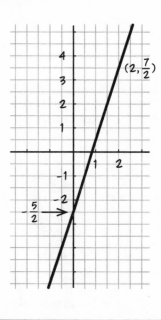

$\left(2, \frac{7}{2}\right)$

From Line to Equation

UP TO NOW, WE'VE STARTED WITH AN EQUATION AND DRAWN ITS GRAPH.

$$y = x + 2$$

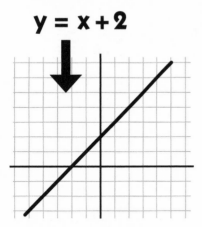

NOW LET'S GO THE OTHER WAY, FROM LINE TO EQUATION. GIVEN A LINE, CAN WE WRITE AN EQUATION WHOSE GRAPH IT IS?

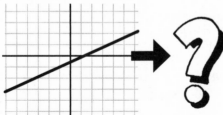

HOW MUCH DO WE NEED TO KNOW ABOUT A LINE TO WRITE ITS EQUATION?

HOW MUCH IS THERE TO KNOW??

KNOWING THE SLOPE ALONE, FOR EXAMPLE, IS NOT ENOUGH. THERE ARE HEAPS OF LINES WITH THE SAME SLOPE. HOW WOULD WE KNOW IF AN EQUATION HAD "OUR" LINE AS ITS GRAPH?

WHOA! $y = mx$ PLUS SOMETHING-OR-OTHER!

ON THE OTHER HAND, IF WE KNOW THE LINE'S SLOPE **AND** y-INTERCEPT, WE CAN WRITE ITS EQUATION IMMEDIATELY—IN SLOPE-INTERCEPT FORM, OF COURSE! FOR EXAMPLE, IF WE KNOW A LINE HAS SLOPE −1 AND y-INTERCEPT −5, ITS EQUATION SIMPLY **MUST** BE $y = -x - 5$.

$$y = \underset{\text{SLOPE}}{-x} \underset{\text{y-INTERCEPT}}{-5}$$

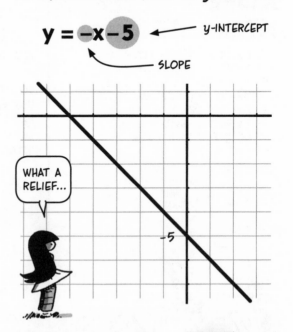

WHAT A RELIEF...

WE CAN DESCRIBE LINES IN SEVERAL DIFFERENT WAYS THAT LEAD TO EQUATIONS.

Point and Slope

THERE'S REALLY NOTHING SPECIAL ABOUT THE y-INTERCEPT. GIVEN **ANY** POINT (a, b), THERE CAN BE ONLY ONE LINE PASSING THROUGH (a, b) WITH A GIVEN SLOPE m.

DIFFERENT LINES, DIFFERENT SLOPES!

THIS LINE'S EQUATION IS

$$y - b = m(x - a)$$

THIS IS CALLED THE EQUATION'S **POINT-SLOPE FORM**. TO SEE THAT THE GRAPH REALLY DOES PASS THROUGH (a, b), WE SOLVE FOR y WHEN $x = a$. PLUGGING IN a FOR x GIVES

$$y - b = m(a - a) = m \cdot 0 \quad \text{SO}$$
$$y - b = 0$$
$$y = b$$

THAT IS, IF $x = a$, THEN $y = b$, SO THE POINT (a, b) IS ON THE EQUATION'S GRAPH.

THE GRAPH HAS SLOPE m, AS YOU CAN SEE BY EXPANDING THE EXPRESSION AND COLLECTING TERMS:

$$y - b = m(x - a)$$
$$y - b = mx - ma$$
$$y = mx + (b - ma)$$

$b - ma$ IS A CONSTANT, SO THIS EQUATION IS IN SLOPE-INTERCEPT FORM, WITH SLOPE m AND y-INTERCEPT $b - ma$.

Example. FIND THE EQUATION OF A LINE PASSING THROUGH THE POINT $(7, 11)$ WITH SLOPE 6.

ANSWER: WE APPLY THE POINT-SLOPE FORMULA DIRECTLY AND GET

$$y - 11 = 6(x - 7)$$

WHICH CAN BE EXPANDED TO FIND THE SLOPE-INTERCEPT FORM:

$$y - 11 = 6x - 42$$
$$y = 6x - 31$$

THE y-INTERCEPT IS -31.

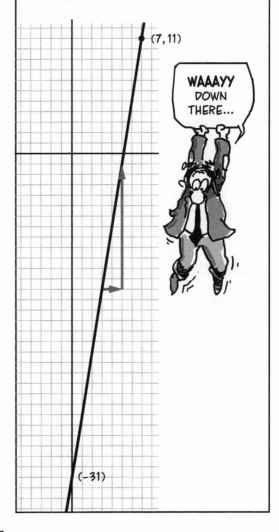

WAAAYY DOWN THERE...

Two Points

GIVEN TWO POINTS, YOU CAN DRAW ONE AND ONLY ONE STRAIGHT LINE THROUGH THEM. ON THE COORDINATE PLANE, IF A LINE PASSES THROUGH TWO POINTS WITH KNOWN COORDINATES (x_1, y_1) AND (x_2, y_2), WHAT IS ITS EQUATION?

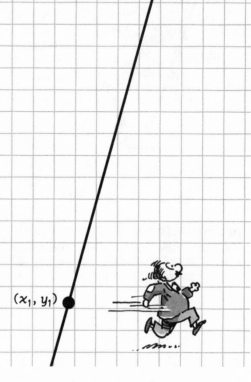

FIRST FIND THE SLOPE m FROM THE DIFFERENCE QUOTIENT.

$$m = \frac{y_2 - y_1}{x_2 - x_1}$$

RISE OVER RUN!

THEN APPLY THE POINT-SLOPE FORMULA USING THIS SLOPE AND EITHER POINT. USING THE FIRST POINT (x_1, y_1), THE EQUATION IS

$$y - y_1 = \left(\frac{y_2 - y_1}{x_2 - x_1}\right)(x - x_1)$$

USING THE SECOND POINT FORMULA GIVES AN EQUATION THAT LOOKS SLIGHTLY DIFFERENT, BUT WORKS OUT TO BE THE SAME.

$$y - y_2 = \left(\frac{y_2 - y_1}{x_2 - x_1}\right)(x - x_2)$$

Example.

FIND THE EQUATION OF THE LINE PASSING THROUGH $(-6, -2)$ AND $(6, 4)$.

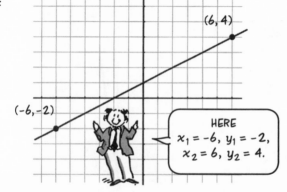

HERE $x_1 = -6$, $y_1 = -2$, $x_2 = 6$, $y_2 = 4$.

ANSWER: FIRST, FORM THE DIFFERENCE QUOTIENT TO FIND THE SLOPE:

$$\frac{4 - (-2)}{6 - (-6)} = \frac{6}{12} = \frac{1}{2}$$

PLUG THIS SLOPE INTO THE POINT-SLOPE FORMULA USING EITHER POINT. WE USE $(6, 4)$.

$$y - 4 = \tfrac{1}{2}(x - 6)$$

$$y - 4 = \tfrac{1}{2}x - 3$$

$$y = \tfrac{1}{2}x + 1$$

Two Equations, Two Lines

LAST CHAPTER, WE LOOKED AT PAIRS OF EQUATIONS IN TWO VARIABLES, LIKE THESE TWO:

$$3x + 4y = 9$$
$$3x + 2y = 6$$

EQUATIONS LIKE THIS, IN THE FORM $ax + by = c$, ARE SAID TO BE IN **STANDARD FORM.** IF $b \neq 0$, THEY CAN EASILY BE REWRITTEN IN SLOPE-INTERCEPT FORM AND GRAPHED:

$$3x + 4y = 9$$
$$4y = -3x + 9$$
$$\boxed{y = -\frac{3}{4}x + \frac{9}{4}}$$

$$3x + 2y = 6$$
$$2y = -3x + 6$$
$$\boxed{y = -\frac{3}{2}x + 3}$$

EQUATIONS OF THE FORM $ax + by = c$ ARE CALLED **LINEAR** EQUATIONS BECAUSE THEIR GRAPHS ARE STRAIGHT LINES.

HEY! CAREFUL WITH THAT THING!

A SOLUTION TO A PAIR OF EQUATIONS IS A PAIR OF NUMBERS (x, y) THAT SATISFIES BOTH EQUATIONS SIMULTANEOUSLY—WHICH MEANS THAT THE POINT (x, y) LIES ON **BOTH THEIR GRAPHS.** IN OTHER WORDS, THE SOLUTION (OR SOLUTIONS) TO A PAIR OF EQUATIONS IS THE POINT (OR POINTS) WHERE THEIR **GRAPHS INTERSECT!!**

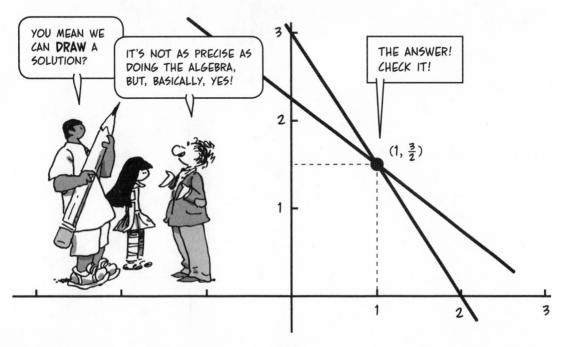

YOU MEAN WE CAN **DRAW** A SOLUTION?

IT'S NOT AS PRECISE AS DOING THE ALGEBRA, BUT, BASICALLY, YES!

THE ANSWER! CHECK IT!

$(1, \frac{3}{2})$

Parallel Lines

AS WE SAW IN THE LAST CHAPTER, A PAIR OF LINEAR EQUATIONS MAY HAVE **NO** SOLUTION... AND NOW WE CAN SEE WHY. THIS CAN HAPPEN ONLY WHEN THE TWO EQUATIONS' GRAPHS **NEVER CROSS.**

AND IF THE LINES **DON'T** CROSS?

LIKE, NEVER?

NEVER EVER?

TWO LINES THAT NEVER CROSS ARE CALLED **PARALLEL LINES,** AND AS YOU CAN SEE, PARALLEL LINES HAVE THE SAME **SLOPE.**

WE CAN EASILY SEE WHETHER A PAIR OF LINEAR EQUATIONS HAVE PARALLEL GRAPHS BY PUTTING THE EQUATIONS INTO SLOPE-INTERCEPT FORM. FOR EXAMPLE:

(1) $3x + 5y = 5$

(2) $6x + 10y = 20$

IN POINT-SLOPE FORM, THESE BECOME

(1a) $y = -\dfrac{3}{5}x + 1$

(2a) $y = -\dfrac{3}{5}x + 2$

THE y-INTERCEPTS ARE DIFFERENT, SO THE GRAPHS ARE TWO SEPARATE LINES, BUT THE SLOPES ARE THE SAME, $-3/5$, SO THE LINES ARE PARALLEL. THE EQUATIONS HAVE NO COMMON SOLUTION.

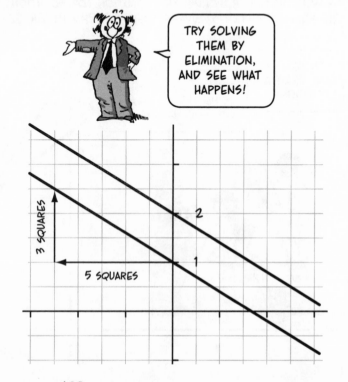

TRY SOLVING THEM BY ELIMINATION, AND SEE WHAT HAPPENS!

3 SQUARES

5 SQUARES

GIVEN TWO LINEAR
EQUATIONS, EITHER:

1. THEY HAVE THE
SAME GRAPH;

2. THEIR GRAPHS
ARE **PARALLEL;**

3. THEIR GRAPHS
CROSS AT A SINGLE
POINT.

ONLY THREE
POSSIBILITIES!

GIVEN TWO EQUATIONS, HOW CAN WE KNOW WHICH OPTION HOLDS TRUE? LET'S SUPPOSE
a, b, c, d, e, AND f ARE SOME FIXED NUMBERS, AND THAT NEITHER b NOR d IS ZERO.
(b AND d WILL BE THE COEFFICIENTS OF y IN THE TWO EQUATIONS, AND WE'LL WANT TO
DIVIDE BY THEM.) WE WANT TO KNOW ABOUT THE STANDARD-FORM EQUATIONS 3 AND 4.

(3) $ax + by = e$

(4) $cx + dy = f$

WE NEXT PUT
THEM IN POINT-
SLOPE FORM:

(3a) $y = -\dfrac{a}{b}x + \dfrac{e}{b}$

(4a) $y = -\dfrac{c}{d}x + \dfrac{f}{d}$

y-INTERCEPTS

SLOPES

COMPARE SLOPES
AND INTERCEPTS,
AND DRAW THIS...

👉 CONCLUSION:

1. IF $a/b = c/d$ **AND** $e/b = f/d$,
THEN THE EQUATIONS' GRAPHS HAVE
THE SAME SLOPE AND y-INTERCEPT. THEY ARE THE
SAME LINE! EVERY POINT ON THIS LINE SOLVES BOTH
EQUATIONS.

2. IF $a/b = c/d$ AND $e/b \neq f/d$, THEN THE GRAPHS
HAVE THE SAME SLOPE BUT DIFFERENT y-INTERCEPTS.
THEY ARE PARALLEL LINES, AND THERE IS NO SOLUTION.

3. IF $a/b \neq c/d$, THEN THE EQUATIONS HAVE DIFFERENT
SLOPES. THEIR GRAPHS CROSS AT A SINGLE POINT,
WHICH SOLVES BOTH EQUATIONS SIMULTANEOUSLY.

IN OTHER WORDS,
IF $a/b = c/d$, DON'T
BOTHER LOOKING
FOR A SOLUTION!

Horizontal and Vertical Lines

IN THE EQUATION

$$ax + by = c$$

WE KEEP ASSUMING $b \neq 0$. THIS MEANS WE'VE LOOKED ONLY AT EQUATIONS LIKE

$$2x + 6y = 4$$

$$9x - 503y = 7,021,077$$

OR EVEN

$$y = 8$$

(a, THE COEFFICIENT OF x, CAN BE ZERO!) BUT WHAT ABOUT WHEN $b = 0$? THESE ARE EQUATIONS LIKE

$$x = c$$

WHICH HAS A **VERTICAL GRAPH.** IT'S ALL THE POINTS WITH AN x-COORDINATE EQUAL TO c. A VERTICAL LINE'S SLOPE IS... **INFINITE.** IT RISES (AND FALLS) FOREVER WITH NO RUN AT ALL.

$(c, 4)$

DOES "FOREVER" MAKE ANYONE ELSE BESIDES ME FEEL... WEIRD...?

$(c, 0)$

$(c, -2)$

$(c, \text{WHATEVER})$

WHEN $a = 0$, ON THE OTHER HAND, THE EQUATION LOOKS LIKE

$$y = c$$

ITS GRAPH IS HORIZONTAL, A LINE OF SLOPE **ZERO** (NO RISE, ALL RUN).

IT DOESN'T RISE, NO MATTER HOW FAR I RUN!

Perpendicular Lines

FINALLY, JUST FOR FUN, LET'S SEE WHAT IT MEANS ALGEBRAICALLY FOR TWO LINES TO MEET AT A RIGHT ANGLE. SUCH LINES ARE CALLED **PERPENDICULAR.** AS YOU GO AROUND THEIR INTERSECTION, ALL FOUR ANGLES ARE EQUAL, LIKE THE CORNERS OF A SQUARE. THE COORDINATE AXES ARE AN EXAMPLE. GIVEN TWO PERPENDICULAR LINES L_1 AND L_2, LET'S SUPPOSE THAT L_1 HAS SLOPE m. WHAT'S THE SLOPE OF L_2 IN TERMS OF m?

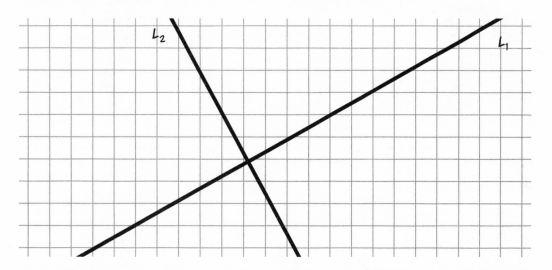

FIRST, SLIDE THE TWO LINES SO THAT THEY CROSS AT THE ORIGIN. THE SLOPES REMAIN THE SAME, AND THE LINE L_1 NOW CONTAINS THE POINT $(1, m)$. (THE LINE $y = mx$ ALWAYS CONTAINS THE POINT $(1, m)$!)

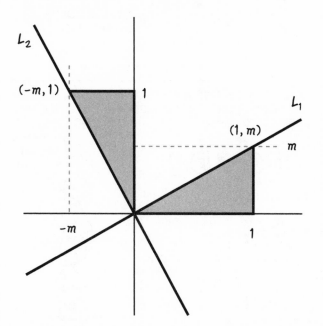

IN THE DIAGRAM, THE TWO GRAY TRIANGLES ARE EXACTLY THE SAME, ONLY TURNED. EACH HAS A SIDE OF LENGTH 1 AND A SIDE OF LENGTH m. SO L_2 CONTAINS THE POINT $(-m, 1)$.

THIS MEANS THAT L_2 HAS SLOPE

$$\frac{1}{-m} = -\frac{1}{m}$$

PERPENDICULAR LINES HAVE SLOPES THAT ARE EACH OTHER'S **NEGATIVE RECIPROCAL** (AS LONG AS NEITHER LINE IS VERTICAL!).

111

WE COVERED (OR UNCOVERED) A LOT IN THIS CHAPTER...

WE BEGAN BY SPRINKLING NUMBERS OVER THE PLANE, SO THAT EACH POINT GETS A UNIQUE PAIR OF COORDINATES (x, y). WE GRAPHED EQUATIONS IN TWO VARIABLES, LIKE $ax + by = c$, AND FOUND THEIR GRAPHS TO BE STRAIGHT LINES.

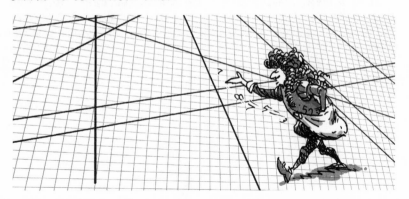

WE LEARNED ABOUT A LINE'S SLOPE, GOING UP AND GOING DOWN.

THE SLOPE CAN EVEN BE INFINITE!

WE DISCOVERED THAT A LINE'S EQUATION IS DETERMINED BY TWO CONDITIONS, WHICH CAN BE ANY OF THESE:

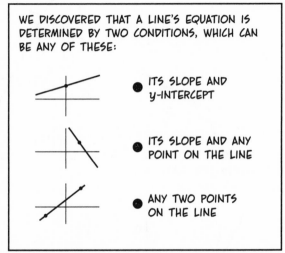

- ITS SLOPE AND y-INTERCEPT

- ITS SLOPE AND ANY POINT ON THE LINE

- ANY TWO POINTS ON THE LINE

WE SAW THAT TWO LINEAR EQUATIONS

$$ax + by = e$$
$$cx + dy = f$$

HAVE A SOLUTION IN COMMON WHEN THEIR GRAPHS CROSS, AND THE SOLUTION (x, y) IS THE CROSSING POINT.

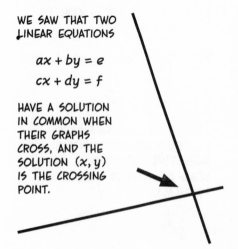

WE ALSO DISCOVERED A TEST FOR PREDICTING WHETHER THE GRAPHS CROSS, NAMELY WHEN $(a/c) \neq (b/d)$. THIS AMOUNTS TO SAYING

$$ad \neq bc$$

SIMPLE!

BECAUSE IF

$$\frac{a}{c} = \frac{b}{d}$$ THEN

$$cd\,\frac{a}{c} = cd\,\frac{b}{d}$$ MULTIPLY-ING BY cd

$$ad = bc$$ CANCELING

I ALSO HOPE TO HAVE PLANTED THE SEED OF THIS IDEA: SOME GRAPHS ARE **NOT** STRAIGHT LINES.

TAKE THIS EQUATION, FOR EXAMPLE: $xy = 1$ OR $y = \dfrac{1}{x}$

ON ITS GRAPH, EACH POINT'S COORDINATES ARE RECIPROCAL TO EACH OTHER. AS LONG AS NEITHER x NOR y IS ZERO, WE CAN MAKE A TABLE OF VALUES AND DRAW THE EQUATION'S GRAPH. IT CURVES!

$y = 1/x$	
x	y
$\frac{1}{5}$	5
$\frac{1}{4}$	4
$\frac{1}{3}$	3
$\frac{1}{2}$	2
1	1
2	$\frac{1}{2}$
3	$\frac{1}{3}$
4	$\frac{1}{4}$
5	$\frac{1}{5}$

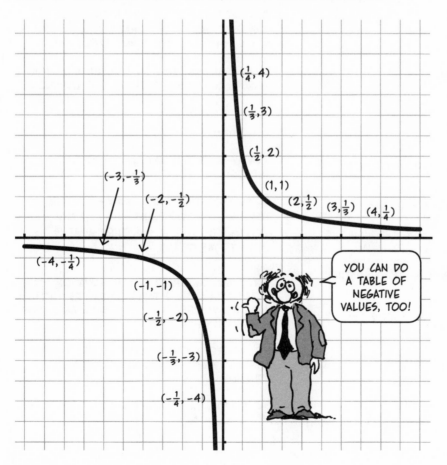

YOU CAN DO A TABLE OF NEGATIVE VALUES, TOO!

BUT LET'S NOT GET AHEAD OF OURSELVES, SHALL WE? FOR NOW, YOU CAN WORK ON SOME PROBLEMS...

113

Problems

MOST OF THESE PROBLEMS REQUIRE GRAPH PAPER. YOU CAN EITHER BUY SOME AT THE STORE, OR DOWNLOAD A PDF FROM HTTP://WWW.LARRYGONICK.COM/TITLES/SCIENCE/THE-CARTOON-GUIDE-TO-ALGEBRA/ AND PRINT AS MANY SHEETS AS YOU NEED.

1. DRAW A SET OF AXES, CHOOSE A UNIT SIZE, AND PLOT THESE POINTS:

$(1,1)$, $(0,6)$, $(-3,0)$, $(-3.5,-0.25)$, $(4,-3)$, $(-4,3)$, $(4,3)$, $(-4,-3)$, $(\frac{1}{2},9)$, $(-\frac{1}{4},-\frac{1}{4})$.

2. DRAW THE GRAPHS OF THESE EQUATIONS:

a. $y = 3x$

b. $y = 3x - 4$

c. $y = -x + 7$

d. $4y = 8 - 2x$

e. $x + y = 5$

f. $2x + 2y = 7$

g. $3x - 2y = 4$

h. $x - 2y = -3$

i. $-3x - 4y = -9$

j. $-14x + 7y = 0$

k. $4y - \frac{1}{2}x = 9$

l. $\frac{x}{2} + \frac{y}{3} = \frac{5}{3}$

m. $4.38 - 1.7y = x$

3. WRITE THE EQUATION OF THE LINE AND DRAW ITS GRAPH.

a. WITH SLOPE 3 AND y-INTERCEPT 5

b. WITH SLOPE 3 PASSING THROUGH THE POINT $(1,1)$

c. WITH SLOPE 500 AND y-INTERCEPT 2,001

d. WITH SLOPE $-\frac{1}{3}$ AND y-INTERCEPT $-\frac{1}{5}$

e. WITH SLOPE -6 PASSING THROUGH THE POINT $(2,3)$

f. WITH SLOPE $\frac{3}{4}$ PASSING THROUGH THE POINT $(-4,-3)$

g. PASSING THROUGH THE POINTS $(-5,-2)$ AND $(-4,1)$

h. PASSING THROUGH THE POINTS $(-2,-2)$ AND $(2,-4)$

4a. DOES THE POINT $(3,4)$ LIE ON THE GRAPH OF THE EQUATION $y = \frac{2}{3}x + 2$? DOES THE POINT $(-3,1)$?

4b. DOES THE POINT $(7,4)$ LIE ON THE LINE WITH SLOPE 2 PASSING THROUGH THE POINT $(5,1)$?

4c. DOES THE POINT $(7,-2)$ LIE ON THE LINE THROUGH THE POINTS $(2,3)$ AND $(3,2)$? WHERE DOES THIS LINE CROSS THE LINE $x = -14$?

5a. WRITE THE EQUATION OF THE LINE PASSING THROUGH THE POINT $(1,2)$ AND PARALLEL TO THE GRAPH OF $8x - 2y = 7$.

5b. WRITE THE EQUATION OF THE LINE PASSING THROUGH THE POINT $(0,3)$ AND PARALLEL TO THE LINE JOINING $(-3,0)$ AND $(3,4)$.

5c. WRITE THE EQUATION OF THE LINE PASSING THROUGH THE POINT $(2.35, 6.147)$ AND PERPENDICULAR TO THE GRAPH OF $y = x$.

5d. WRITE THE EQUATION OF THE LINE PERPENDICULAR TO THE GRAPH OF $y = 5$ AND PASSING THROUGH THE POINT $(700, -31)$.

6. BY GRAPHING, FIND AN APPROXIMATE SIMULTANEOUS SOLUTION TO THESE TWO EQUATIONS.

$$13.408x + 3.2y = 47.82$$
$$1.479x - 1.7y = -2.295$$

7. WHY DOES THE LINE $y = mx$ PASS THROUGH THE POINT $(1, m)$?

8. SUPPOSE THE LINE $y = mx + b$ PASSES THROUGH THE TWO POINTS (x_1, y_1) AND (x_2, y_2). IF $x_2 = x_1 + p$, THEN WHAT IS y_2 IN TERMS OF y_1?

9a. DRAW THE GRAPH OF THE EQUATION $xy = 6$. (START BY MAKING A TABLE OF VALUES, INCLUDING NEGATIVE VALUES.) USING THE SAME AXES, DRAW THE GRAPH OF $x + y = 5$.

9b. WHERE, APPROXIMATELY, DO THE GRAPHS INTERSECT? CAN YOU SOLVE THIS PAIR OF EQUATIONS?

9c. DO THE SAME WITH THE EQUATIONS $xy = 6$ AND $x - y = 5$.

Chapter 9
Power Play

Up to now, we've been careful always to put plus or minus signs between our variables (or their multiples, as in $4x + 2y$). It was only at the very end of the last chapter that we wrote xy, with nothing in between.

HEEHEE HEEHEEHEE!

(TRUE, WE'VE WRITTEN EXPRESSIONS LIKE ax THAT ARE, STRICTLY SPEAKING, THE PRODUCT OF TWO VARIABLES... BUT INWARDLY, WE OFTEN THINK OF a AS A STAND-IN FOR A FIXED NUMBER, WHILE x REALLY, YOU KNOW, VARIES.)

HOLD STILL, YOU SLIPPERY LITTLE WEASEL!

a

x

IN THIS CHAPTER, WE BEGIN TO MULTIPLY VARIABLES TOGETHER, AND TO DIVIDE BY THEM... AND WE'LL ALSO START USING LETTERS LIKE a AND b FOR "REAL," VARYING VARIABLES.

NICE LITTLE a AND b...

UH-OH... NOW THEY'RE WIGGLING TOO!

a b

THE FIRST MULTIPLICATION WILL BE THE PRODUCT OF A VARIABLE TIMES **ITSELF**, AS IN **xx.** MULTIPLY BY x AGAIN TO MAKE xxx, AND REPEAT WITH $xxxx$, $xxxxx$, ETC., MAKING AN x-PARADE AS LONG AS YOU LIKE.

UM... WE'RE RUNNING OUT OF PAGE HERE...

TO SAVE INK AND PAPER, WE RESORT TO SHORTHAND, WRITING x^2 FOR xx, x^3 FOR xxx, x^4 FOR $xxxx$, ETC. WE SAY THAT x IS RAISED TO THE SECOND, THIRD, OR FOURTH **POWER**, AND READ x^4 AS "x TO THE FOURTH." THE EXPRESSION x^n ("x TO THE ENTH") WOULD BE THE PRODUCT OF n FACTORS OF x. THE LITTLE RAISED NUMBER IS CALLED AN **EXPONENT**.

← EXPONENT

Numerical EXAMPLES

$1^2 = 1 \times 1 = \mathbf{1}$

$2^2 = 2 \times 2 = \mathbf{4}$

$2^3 = 2 \times 2 \times 2 = \mathbf{8}$

$(-5)^3 = (-5) \times (-5) \times (-5)$
$= 25 \times (-5) = \mathbf{-125}$

$(-8)^2 = (-8)(-8) = \mathbf{64}$

$(1.5)^5 = (1.5 \times 1.5) \times (1.5 \times 1.5) \times (1.5)$
$= (2.25) \times (2.25) \times 1.5$
$= 5.0625 \times 1.5$
$= \mathbf{7.59375}$

AND BY THE WAY, $x^1 = x$, WITH ONLY ONE "FACTOR."

HEY, WHEN ARE YOU GOING TO START USING a INSTEAD OF x?

116

THE LOW POWERS x^2 AND x^3 HAVE SPECIAL NAMES. x^2 IS **"x SQUARED"** BECAUSE IT'S THE AREA OF A SQUARE WITH ALL SIDES EQUAL TO x.

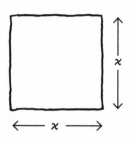

x^3 IS **"x CUBED"**: IT'S THE VOLUME OF A CUBE WITH ALL SIDES EQUAL TO x.

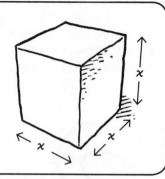

HERE'S A TABLE OF SEVERAL SQUARES AND CUBES. YOU CAN SEE THAT SQUARES ARE NEVER NEGATIVE; x^2 IS THE PRODUCT OF TWO NUMBERS OF THE SAME SIGN, AS IN $(-5)(-5) = 25$. CUBES OF NEGATIVE NUMBERS, THOUGH, ARE ALWAYS NEGATIVE. $(-5)(-5)(-5) = (25)(-5) = -125$. (SEE PAGE 53.)

THREE MINUS SIGNS MAKE A MINUS?

YES, ISN'T IT ODD? (THE NUMBER THREE, I MEAN...)

x	x^2	x^3
-6	36	-216
-5	25	-125
-4	16	-64
-3	9	-27
-2	4	-8
-1	1	-1
0	0	0
1	1	1
2	4	8
3	9	27
4	16	64
5	25	125
6	36	216

NOW WE CAN WRITE A NEW KIND OF ALGEBRAIC EXPRESSION, ANYTHING LIKE THIS:

WAIT—WHAT DO YOU DO FIRST?

YEAH... IS THAT $(3x)^2$ OR $3(x^2)$?

USING THIS NEW OPERATION, RAISING TO A POWER OR **EXPONENTIATION,** IN AN EXPRESSION BRINGS UP THAT AGE-OLD PROBLEM, "WHAT'S THE ORDER?" THE RULE ON PAGE 38 (MULTIPLY BEFORE ADDING) MUST BE EXTENDED TO EXPONENTS. THE EXTENDED RULE IS: IN THE ABSENCE OF PARENTHESES, ALWAYS EVALUATE **POWERS BEFORE MULTIPLICATIONS** (OR DIVISIONS).

MEANS

1. SQUARE x
2. MULTIPLY BY 3

IN OTHER WORDS, $3(x^2)$!

Examples

1. EVALUATE $3 \cdot 4^2 + 9$

THE RULE: FIRST THE EXPONENT, THEN THE PRODUCT, THEN THE SUM.

$$3 \cdot 4^2 + 9 = 3 \cdot 16 + 9$$
$$= 48 + 9 = 57$$

2. EVALUATE $ab^3 - 18$ WHEN $a = 3$, $b = 2$

FIRST, PLUG IN THE VALUES TO GET

$$3 \cdot 2^3 - 18$$

EVALUATE THE CUBE 2^3 BEFORE DOING THE REST.

$$3 \cdot 2^3 - 18 = 3 \cdot 8 - 18$$
$$= 24 - 18$$
$$= 6$$

EXPONENTS DON'T REALLY GIVE TOO MUCH TROUBLE, DO THEY?

QUITE RIGHT... THEY'RE VERY GOOD LITTLE NUMBERS...

BECAUSE THEY OBEY!

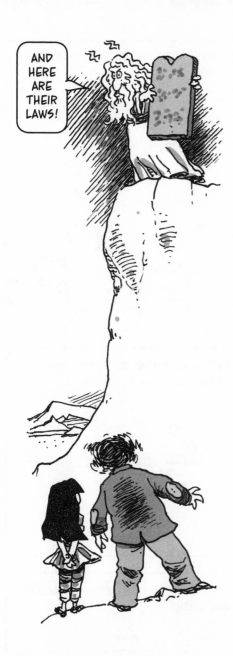

AND HERE ARE THEIR LAWS!

LAWS of EXPONENTS:

IN THESE LAWS, a AND b CAN BE ANY NUMBERS, WHILE m AND n ARE POSITIVE INTEGERS.

1. $a^n a^m = a^{(n+m)}$

$a^n a^m$ IS THE PRODUCT OF n FACTORS OF a MULTIPLIED BY m FACTORS OF a,

$$\underbrace{a \cdot a \cdot \ldots \cdot a}_{n} \cdot \underbrace{a \cdot a \cdot \ldots \cdot a}_{m}$$

MAKING A TOTAL OF $n+m$ FACTORS.

2. $(a^n)^m = a^{nm}$

WE CAN WRITE THE PRODUCT $(a^n)^m$ THIS WAY:

$$\left. \begin{array}{c} a \cdot a \cdot \ldots \cdot a \\ \cdot \, a \cdot a \cdot \ldots \cdot a \\ \vdots \\ \cdot \, a \cdot a \cdot \ldots \cdot a \end{array} \right\} m \text{ ROWS}$$

$\underbrace{\qquad\qquad}$
n FACTORS
IN EACH ROW

THERE ARE nm FACTORS ALTOGETHER.

3. $(ab)^n = a^n b^n$

THIS COMES FROM THE COMMUTATIVE LAW.

$$(ab)^n = ab \cdot ab \cdot \ldots \cdot ab$$

REARRANGING (OR "COMMUTING") THE ORDER GIVES

$$a \cdot a \cdot \ldots \cdot a \cdot b \cdot b \cdot \ldots \cdot b = a^n b^n$$

Examples

$$3^2 3^3 = 3^{2+5} = 3^5 = \mathbf{243}$$

$$(a^2 b)^3 = (a^2)^3 b^3 = \mathbf{a^6 b^3}$$

$$(2t^2 u)^2 = \mathbf{4t^4 u^2}$$

119

Powers "DOWN THERE"

LET'S INVERT ONE OF THESE POWERS BY PUTTING IT IN THE DENOMINATOR.

GETTIN' **DOWN!**

$$\frac{1}{a^2}$$

NOW MULTIPLY THAT EXPRESSION BY a^3.

$$a^3 \frac{1}{a^2} = \frac{a^3}{a^2} = \frac{aaa}{aa}$$

$$= \left(\frac{a}{a}\right)\left(\frac{a}{a}\right)\frac{a}{1}$$

$$= a$$

REMEMBER, $a/a = 1$!

IN OTHER WORDS, JUST AS WITH NUMERIC FRACTIONS, A FACTOR COMMON TO BOTH NUMERATOR AND DENOMINATOR **CANCELS OUT.** WE CAN WRITE IT OUT LIKE THIS:

$$\frac{a^3}{a^2} = \frac{\cancel{a}\cancel{a}a}{\cancel{a}\cancel{a}} = a$$

OR MORE SIMPLY,

$$\frac{a^3}{a^2} = a$$

THIS FACT GIVES US A NICE FORMULA FOR **ANY** EXPONENTS. IF n AND m ARE POSITIVE INTEGERS WITH $n > m$, AND a IS NONZERO, THEN

$$\frac{a^n}{a^m} = a^{n-m}$$

WHICH IS TRUE BECAUSE EXACTLY m FACTORS OF a CANCEL OUT.

$$\frac{\overbrace{\cancel{a} \cdot \cancel{a} \cdot \ldots \cdot \cancel{a}}^{m} \cdot \overbrace{a \cdot \ldots \cdot a}^{n-m}}{\underbrace{\cancel{a} \cdot \cancel{a} \cdot \ldots \cdot \cancel{a}}_{m}}$$

$n-m$ IS THE NUMBER OF FACTORS LEFT OVER!

I LOVE LEFTOVERS!

EXPONENTS CAN ALSO BE
ZERO or LESS!

WE JUST SAW THAT $a^n/a^m = a^{n-m}$. THIS WOULD MEAN, WHEN $m = n$, THAT

$$\frac{a^n}{a^n} = a^{n-n} = a^0$$

BUT OF COURSE a^n/a^n IS ALSO = 1, AS THE RATIO OF A NUMBER TO ITSELF. THAT'S WHY WE **DEFINE** a^0 BY

$$a^0 = 1$$

NO MATTER WHAT a IS (EXCEPT ZERO). $3^0 = 6^0 = (-156.71)^0 = 1$. 0^0 IS BEST LEFT UNDEFINED.

AND NEGATIVE EXPONENTS? WHAT'S a^{-n} TO MEAN? IF WE INSIST THAT THE EXPONENT LAWS APPLY, THEN THIS WOULD BE TRUE:

$$a^n a^{-m} = a^{n-m}$$

WHICH IS SO ONLY IF

$$a^{-m} = \frac{1}{a^m}$$

AND THAT'S HOW WE DEFINE a^{-m}.

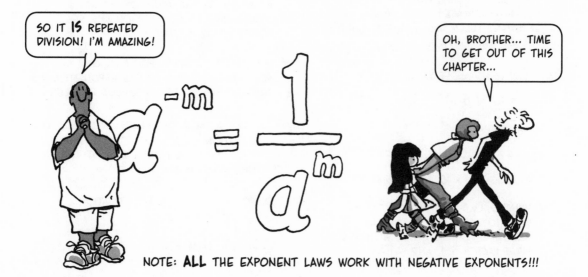

NOTE: **ALL** THE EXPONENT LAWS WORK WITH NEGATIVE EXPONENTS!!!

Problems

1. EVALUATE:

a. 2^1

b. 2^2

c. 2^3

d. 2^{-4}

e. 2^{-5}

f. 2^6

g. $(-2)^6$

h. $(-3)^4$

i. $5^2 5^3$

j. $2^2 \cdot 4^2$

k. $(2 \cdot 4)^2$

l. $-3 \cdot 2^5 - 100$

m. $3^3 3^{-2} + 6^2 (3-1)^{-1}$

n. 3^{-3}

o. $(1/3)^{-3}$

p. $(3/5)^{-1}$

q. $(10^{-3})^2$

r. $3^2 - 3^{-2}$

s. $5x^2$ WHEN $x = 3$

t. $x^2 + x + 1$ WHEN $x = 1$

u. $x^2 + x + 1$ WHEN $x = 2$

v. $x^2 + x + 1$ WHEN $x = 3$

w. $a^2 x + ax^2$ WHEN $a = 2$, $x = 3$.

2. IS $(-6)^{100}$ POSITIVE OR NEGATIVE? HOW ABOUT -6^{100}? HOW ABOUT $(-6)^{-100}$?

3. WHAT IS $\dfrac{3^{101}}{3^{100}}$?

4. SIMPLIFY:

a. $p^4 p^3$

b. $t(5t^2)$

c. $6x^{-4} x^9$

d. $4^{-2} u^{-2} u^{-1}$

e. $(3x^2)^3$

f. $(2x^3)^2$

g. $(-a^2 x)^3$

h. $(a^2 b^{-2})^2$

i. $a^7 b a^3 b^4$

j. $(a^{-1})^n$

k. $\dfrac{2x}{(4x)^{-2}}$

5. WHAT IS $t^n \left(\dfrac{1}{t}\right)^n$?

6. WHAT IS 10^2? 10^3? 10^4? 10^5? 10^6? HOW MANY ZEROES AFTER THE INITIAL 1 ARE THERE IN 10^{25}?

WE CAN WRITE VERY LARGE (AND VERY SMALL) NUMBERS IN WHAT IS CALLED "SCIENTIFIC NOTATION" BY USING POWERS OF 10. FOR EXAMPLE, WE CAN WRITE

$$3,150,000 = 3.15 \times 10^6$$
$$57,830 = 5.783 \times 10^4.$$

IN SCIENTIFIC NOTATION, THE FIRST FACTOR HAS ONE DIGIT TO THE LEFT OF THE DECIMAL POINT, AND THE SECOND FACTOR IS A POWER OF 10 WHOSE EXPONENT TELLS YOU HOW MANY DIGITS FOLLOW THE FIRST.

7a. SHOW USING ALGEBRA THAT

$$a \cdot 10^n + b \cdot 10^n = (a+b) \cdot 10^n$$

b. SHOW THAT

$$(a \times 10^n)(b \times 10^m) = ab \times 10^{n+m}$$

c. WHAT IS $(3.1 \times 10^{15}) + (2.5 \times 10^{15})$?

d. WHAT IS $(3.5 \times 10^4)(3 \times 10^8)$? BE SURE TO WRITE THE ANSWER IN SCIENTIFIC NOTATION, I.E., WITH THE FIRST FACTOR ≥ 1 AND < 10.

8. EVALUATE $\dfrac{x^2 + 2x + 1}{x + 1}$ FOR SEVERAL DIFFERENT VALUES OF x (NOT $x = -1$, THOUGH!). DO YOU NOTICE ANYTHING INTERESTING ABOUT THE ANSWERS IN RELATION TO THE VALUE OF x? WRITE YOUR GUESS AS AN EQUATION AND SEE WHERE IT LEADS YOU.

9. WHAT IS 2^{12}? (HINT: FOR A QUICK SOLUTION, USE ONE OF THE EXPONENT LAWS AND THE RESULT OF A PREVIOUS PROBLEM.)

10a. MAKE A TABLE OF VALUES (x, y) FOR THE EQUATION $y = x^2$ (OR COPY IT FROM PAGE 117). DRAW THE GRAPH, AS BEST YOU CAN, OF THE EQUATION $y = x^2$.

b. DO THE SAME FOR THE EQUATION $y = x^3$.

c. DO THE SAME FOR $y = x^2 - 2x + 1$.

Chapter 10
Rational Expressions

WE'RE NOW READY TO DIVIDE BY ENTIRE EXPRESSIONS, NOT JUST INDIVIDUAL VARIABLES. THE RESULT IS SOMETHING CALLED A **RATIONAL** EXPRESSION: IT'S THE RATIO BETWEEN AN ALGEBRAIC NUMERATOR AND AN ALGEBRAIC DENOMINATOR—IN OTHER WORDS, AN EXPRESSION OVER AN EXPRESSION!

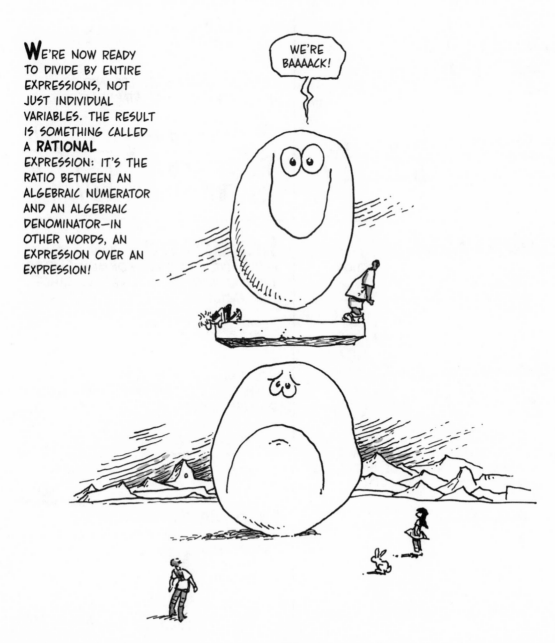

MULTIPLYING
Rational Expressions

IS AS EASY AS MULTIPLYING FRACTIONS. IT'S TOP TIMES TOP AND BOTTOM TIMES BOTTOM, LIKE THIS:

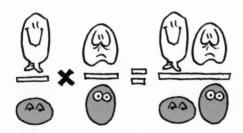

OR, IF YOU PREFER LETTERS—IF a, b, c, AND d ARE ANY EXPRESSIONS, THEN

$$\frac{a}{b}\frac{c}{d} = \frac{ac}{bd}$$

A RATIONAL EXPRESSION'S **RECIPROCAL**, LIKE A FRACTION'S, IS ITS INVERSE, WHAT YOU GET BY TURNING THE EXPRESSION UPSIDE DOWN. (REMEMBER, x^{-1} IS THE RECIPROCAL OF x.)

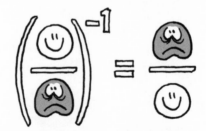

OR, IF YOU PREFER LETTERS, AND BY NOW YOU'RE PROBABLY STARTING TO...

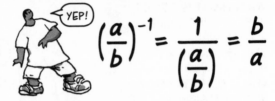

YEP!

$$\left(\frac{a}{b}\right)^{-1} = \frac{1}{\left(\frac{a}{b}\right)} = \frac{b}{a}$$

Dividing BY A RATIONAL EXPRESSION IS LIKE DIVIDING BY ANY FRACTION.

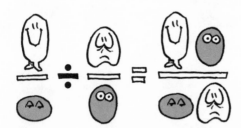

BECAUSE DIVIDING IS THE SAME AS MULTIPLYING BY THE RECIPROCAL, WE INVERT THE DIVISOR AND MULTIPLY. WITH LETTERS,

$$\frac{a}{b} \div \frac{c}{d} = \frac{a}{b} \cdot \frac{d}{c}$$
$$= \frac{ad}{bc}$$

Important: IF NUMERATOR AND DENOMINATOR HAVE ANY **COMMON FACTOR**, THE MULTIPLICATION RULE LETS US **CANCEL** THAT FACTOR.

$$\frac{ac}{ad} = \frac{a}{a} \cdot \frac{c}{d} = 1 \cdot \frac{c}{d} = \frac{c}{d}$$

HERE THE COMMON FACTOR a CANCELED OUT, AND WE CAN SIMPLY WRITE

$$\frac{\cancel{a}c}{\cancel{a}d} = \frac{c}{d}$$

OUT! OUT!

Example 1.

$$\frac{a^2ct^2}{5x} \cdot \frac{10x^3}{ac^2} = \frac{\overset{2}{\cancel{10}}a^2\cancel{c}t^2x^{\overset{2}{\cancel{3}}}}{\cancel{5}\cancel{a}c^{\cancel{2}}\cancel{x}}$$

$$= \frac{2at^2x^2}{c} \quad \text{AFTER ALL CANCELLATIONS.}$$

ADDING Rational

Expressions, LIKE ADDING FRACTIONS, CAN BE A PAIN IN THE SIT-BONES. LUCKILY, FRACTIONS AND EXPRESSIONS ADD IN EXACTLY THE SAME WAY, SO YOU SHOULD BE ON FAMILIAR GROUND.

ADD FRACTIONS? NO PROBLEM! I JUST WALK THE OTHER WAY...

YOU'LL GO FAR, IN SOME DIRECTION...

SOMETIMES ADDING FRACTIONS IS EASY. WHEN THE SUMMANDS (THE FRACTIONS TO BE ADDED) ALL HAVE THE SAME DENOMINATOR, SIMPLY ADD THE NUMERATORS, KEEP THE DENOMINATOR, AND YOU'RE DONE.

OH... YEAH... THAT'S NOT **SO** BAD...

WITH RATIONAL EXPRESSIONS, IT'S THE SAME. IF THEY HAVE THE SAME DENOMINATOR, JUST ADD THE NUMERATORS AND KEEP THE DENOMINATOR.

$$(7) \qquad \frac{m}{d} + \frac{n}{d} = \frac{m+n}{d}$$

Y'KNOW, I THINK I CAN LIVE WITH THAT...

IN CASE YOU'RE WONDERING WHY ALL THESE FORMULAS ARE TRUE FOR EXPRESSIONS—IT'S BECAUSE THE SAME FORMULAS ARE TRUE FOR **NUMBERS!** AFTER ALL, AN EXPRESSION IS NOTHING BUT A NUMBER WAITING TO BE EVALUATED!!!

THAT'S ONE WAY TO LOOK AT ME!

Example 2.

$$\frac{a}{x^2y^2z^2} + \frac{1}{x^2y^2z^2} = \frac{a+1}{x^2y^2z^2}$$

125

Adding Expressions with
DIFFERENT DENOMINATORS

THE FUN WITH ADDITION BEGINS WHEN EXPRESSIONS HAVE DIFFERENT DENOMINATORS. THEN, AS WITH FRACTIONS, WE MUST FIND A **COMMON** OR SHARED DENOMINATOR. FOR EXAMPLE, TO ADD 1/3 + 1/5, WE FIRST EXPRESS BOTH FRACTIONS AS **FIFTEENTHS,** 15 BEING THE PRODUCT OF THE DENOMINATORS, 3 × 5.*

$$\frac{1}{5} = \frac{3 \times 1}{3 \times 5} = \frac{3}{15}$$

$$\frac{1}{3} = \frac{5 \times 1}{5 \times 3} = \frac{5}{15}$$

$$\frac{3}{15} + \frac{5}{15} = \frac{8}{15}$$

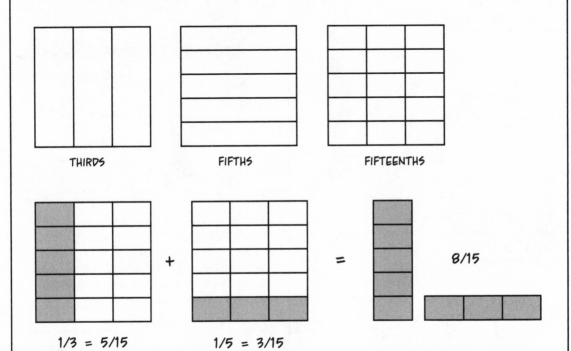

THIRDS   FIFTHS   FIFTEENTHS

1/3 = 5/15   1/5 = 3/15   8/15

WE CAN ALWAYS MAKE A COMMON DENOMINATOR OF TWO FRACTIONS BY MULTIPLYING THEIR DENOMINATORS—AND THE SAME GOES FOR RATIONAL EXPRESSIONS.

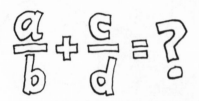

SADLY, I DON'T HAVE A DIAGRAM FOR EXPRESSIONS LIKE THE ONE FOR NUMBERS. THIS IS PURE ALGEBRA!

JUST DO IT!

*IF THIS DOESN'T RING A BELL, YOU SHOULD GO BACK AND REFRESH YOUR ARITHMETIC.

TO ADD

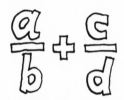

WE FOLLOW THE SAME STEPS AS WE DID WITH THIRDS AND FIFTHS. THE PRODUCT **bd** IS A COMMON DENOMINATOR, BECAUSE

HERE WE MULTIPLIED BY d/d, WHICH IS EQUAL TO 1...

AND HERE BY b/b!

NOW WE HAVE TWO EXPRESSIONS WITH THE SAME DENOMINATOR bd, AND THESE CAN BE ADDED AS ON PAGE 125. RESULT:

EQUATION 8:

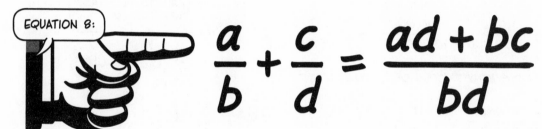

$$\frac{a}{b} + \frac{c}{d} = \frac{ad + bc}{bd}$$

Example 3.

$$\frac{2}{x} + \frac{3}{y} = \frac{2y + 3x}{xy}$$

MULTIPLY ACROSS, LIKE THIS!

NOTE!!

WHEN $b = 1$, THAT IS, WHEN ONE TERM IS "ALL NUMERATOR," EQUATION 8 SAYS THIS:

$$a + \frac{c}{d} = \frac{ad + c}{d}$$

WHICH SHOWS HOW TO ADD THE NUMBER 1 (OR ANY CONSTANT) TO A RATIONAL EXPRESSION.

$$1 + \frac{P}{Q} = \frac{Q + P}{Q}$$

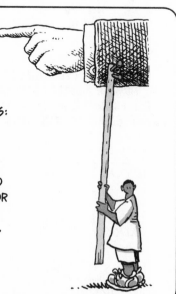

Different Denominators (continued)

IT IS ALWAYS POSSIBLE TO FIND A COMMON DENOMINATOR BY MULTIPLYING THE DENOMINATORS OF THE SUMMANDS. ALAS, THIS PRODUCT MAY BE WAY TOO BIG. WE WANT TO AVOID BIG, HAIRY DENOMINATORS WHENEVER POSSIBLE.

BIG, HAIRY DENOMINATOR, BEST AVOIDED

FOR EXAMPLE, TRY ADDING $\frac{1}{10,000} + \frac{1}{1,000}$. THE DENOMINATORS' PRODUCT IS A WHOPPING 10,000,000... AND THE SUM WORKS OUT TO

$$\frac{1,000}{10,000,000} + \frac{10,000}{10,000,000} = \frac{11,000}{10,000,000}$$

I'M MORE TO BE PITIED THAN FEARED...

WHICH HAS MANY FACTORS OF 10 TO CANCEL:

$$\frac{11,\cancel{000}}{10,000,\cancel{000}} = \frac{11}{10,000}$$

THE FINAL DENOMINATOR IS MUCH SMALLER, AND WE CONCLUDE THAT 10 MILLION WAS UNNECESSARILY BIG AND HAIRY.

IT WOULD HAVE BEEN BETTER TO FIND A SMALLER COMMON DENOMINATOR IN THE FIRST PLACE. THIS NUMBER MUST BE A MULTIPLE OF BOTH 1,000 AND 10,000... AND WE SEE THAT 10,000 ITSELF FILLS THE BILL. IT'S 1×10,000 AND 10×1,000, A MULTIPLE OF BOTH. IN FACT, IT'S THE **LEAST COMMON MULTIPLE** OF THE ORIGINAL DENOMINATORS, AND IT WORKS PERFECTLY.

$$\frac{1}{10,000} + \frac{1}{1,000}$$
$$=$$
$$\frac{1}{10,000} + \frac{10}{10,000}$$
$$=$$
$$\frac{11}{10,000}$$

NO CANCELLATION NECESSARY!

NOW LET'S TRY AN ALGEBRAIC SUM AND SEE IF WE CAN SHAVE DOWN ITS DENOMINATOR. START WITH THIS:

$$\frac{1}{a} + \frac{1}{a^2}$$

I'VE ALWAYS HAD A THING FOR BALD DENOMINATORS...

WE CAN USE WHAT WE'VE ALREADY LEARNED FROM EQUATION 8, MAKING A COMMON DENOMINATOR BY MULTIPLYING a TIMES a^2 TO GET a^3. THEN

$$\frac{1}{a} = \frac{a^2}{a^2}\frac{1}{a} = \frac{a^2}{a^3}$$

$$\frac{1}{a^2} = \frac{a}{a}\frac{1}{a^2} = \frac{a}{a^3}$$

ADDING THESE GIVES

$$\frac{1}{a} + \frac{1}{a^2} = \frac{a^2}{a^3} + \frac{a}{a^3}$$
$$= \frac{a^2 + a}{a^3}$$

HMM... a IS IN **EVERYTHING**... IT JUST **HAS** TO CANCEL...

THE NUMERATOR HAS A FACTOR OF a, BECAUSE, BY THE DISTRIBUTIVE LAW,

$$a^2 + a = a(a + 1)$$

THIS CANCELS ONE FACTOR OF a DOWNSTAIRS, AND WE GET

$$\frac{a(a+1)}{a^{3^2}} = \frac{a + 1}{a^2}$$

THE FINAL DENOMINATOR, a^2, HAS A LOWER POWER THAN a^3... AND WE HAVE TO THINK THAT a^3 WAS JUST A LITTLE HAIRIER THAN IT NEEDED TO BE...

RAZOR!

CAN WE ADD $1/a$ AND $1/a^2$ WITHOUT THE FINAL CANCELLATION? IS THERE A LESS HAIRY COMMON DENOMINATOR THAN a^3? IF SO, THIS DENOMINATOR MUST BE A MULTIPLE OF BOTH a AND a^2, BUT SOMEHOW SIMPLER AND BETTER THAN a^3...

HOLD VERY STILL NOW...

YES, THE DENOMINATOR WE WANT IS NOTHING OTHER THAN...

I JUST TOOK A LITTLE OFF THE EXPONENT!

WE SEE THAT a^2 IS A MULTIPLE OF a:

$$a^2 = a \cdot a$$

AND IT'S OBVIOUSLY A MULTIPLE OF ITSELF!

$$a^2 = 1 \cdot a^2$$

NOW WE CAN EXPRESS EACH TERM OF THE SUM WITH THE DENOMINATOR a^2:

$$\frac{1}{a} + \frac{1}{a^2} = \frac{a \cdot 1}{a \cdot a} + \frac{1}{a^2}$$

$$= \frac{a}{a^2} + \frac{1}{a^2}$$

IF YOU DON'T GET THE SAME ANSWER AS BEFORE, I'LL NEVER BELIEVE YOU AGAIN...

$$= \frac{a+1}{a^2}$$

PHEW!

IN GENERAL, GIVEN TWO POWERS OF A VARIABLE, THEIR **LEAST COMMON MULTIPLE IS SIMPLY THE GREATER POWER.** t^5 IS A MULTIPLE OF t^2, x^{98} IS A MULTIPLE OF x^{97}, AND SO FORTH.

HERE IS A SUM WITH DENOMINATORS CONTAINING SEVERAL DIFFERENT FACTORS.

$$\frac{2p}{x^3yz^{10}} + \frac{x+3}{x^2y^5z}$$

MAN, THIS PROBLEM IS PRETTY HAIRY ALREADY!

IT HAPPENS...

IT LOOKS PRETTY HORRIBLE, BUT AFTER THE LAST PAGE, WE KNOW WHAT TO DO. TO FIND A LEAST COMMON MULTIPLE (LCM) OF THE DENOMINATORS, **FIND THE HIGHEST POWER OF EACH VARIABLE IN THE DENOMINATORS, AND MULTIPLY THESE POWERS TOGETHER.**

THE DENOMINATORS' VARIABLES ARE x, y, AND z (p IS ONLY IN THE NUMERATOR!). THE LARGEST EXPONENT OR POWER OF x IS 3; THE LARGEST EXPONENT OF y IS 5; THE LARGEST EXPONENT OF z IS 10. SO **THE LCM IS $x^3y^5z^{10}$.**

$$x^3y^5z^{10} = y^4(x^3yz^{10})$$

THE FIRST DENOMINATOR

$$= xz^9(x^2y^5z)$$

THE SECOND DENOMINATOR

HERE YOU SEE THAT IT IS A MULTIPLE OF BOTH DENOMINATORS.

THE SUM NOW LOOKS LIKE

$$\frac{y^4}{y^4} \cdot \frac{2p}{x^3yz^{10}} + \frac{xz^9}{xz^9} \cdot \frac{(x+3)}{x^2y^5z}$$

$$= \frac{2py^4}{x^3y^5z^{10}} + \frac{x^2z^9 + 3xz^9}{x^3y^5z^{10}}$$

$$= \frac{2py^4 + x^2z^9 + 3xz^9}{x^3y^5z^{10}}$$

WHOA! YOU COULDN'T GIVE THIS MORE OF A SHAVE?

HEY, IT COULD BE HAIRIER!!!!

Example 4 (MUCH SIMPLER!). ADD $\frac{1}{a} + \frac{1}{ab}$.

HIGHEST POWER OF a IS 1; HIGHEST POWER OF b, DITTO; SO DENOMINATORS' LCM IS SIMPLY ab, AND THE SUM BECOMES

$$\frac{1}{a} + \frac{1}{ab} = \frac{b}{ab} + \frac{1}{ab} = \frac{b+1}{ab}$$

LUCKILY, MOST OF OUR PROBLEMS WILL LOOK MORE LIKE THIS THAN THAT HAIRY ONE!

GOOD.

A FEW MORE THINGS
to Think About

WHEN WHOLE NUMBERS APPEAR AS FACTORS IN DENOMINATORS, THEY MUST BE TAKEN INTO ACCOUNT WHEN FINDING THE DENOMINATORS' LCM.

OH, YEAH... NUMBERS... I FORGOT ABOUT THEM...

Example 5. FIND

$$\frac{b^2}{8a^2} - \frac{5}{6ab}$$

THE LCM OF 8 AND 6 IS 24; THE LCM OF a^2 AND ab IS a^2b; SO THE LCM OF THE DENOMINATORS IS $24a^2b$. WITH THIS COMMON DENOMINATOR, WE ADD:

$$\frac{(3b)\cdot b^2}{(3b)8a^2} - \frac{(4a)(5)}{(4a)6ab}$$

$$= \frac{3b^3}{24a^2b} - \frac{20a}{24a^2b}$$

$$= \frac{3b^3 - 20a}{24a^2b}$$

IF WE HAD SIMPLE-MINDEDLY MULTIPLIED THE DENOMINATORS, WE WOULD HAVE BEEN STUCK WITH THIS COMMON DENOMINATOR INSTEAD!

I CAN'T DECIDE WHETHER TO MAJOR IN MATH OR GO TO BARBER COLLEGE...

FINALLY, REMEMBER THAT THE LETTER VARIABLES a, b, c, ETC., CAN STAND FOR ENTIRE **EXPRESSIONS.** ALL THE FORMULAS IN THIS CHAPTER ARE TRUE FOR EXPRESSIONS IN PLACE OF THE VARIABLES. EQUATION 8, FOR INSTANCE, MIGHT LOOK LIKE THIS:

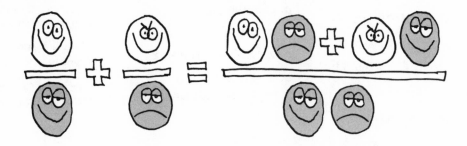

Example 6. ADD

$$\frac{1}{(x+1)(x+2)^2} + \frac{1}{(x+1)^2(x+2)}$$

IN THIS EXAMPLE, WE TREAT THE EXPRESSIONS $x+1$ AND $x+2$ **AS IF THEY WERE VARIABLES.** (IN FACT, THEY ARE VARIABLES. $x+1$ COULD BE CALLED a, AND $x+2$ COULD BE CALLED b.) THEN WE PERFORM THE ADDITION EXACTLY AS BEFORE.

THE HIGHEST POWER OF $x+1$ IS 2; THE HIGHEST POWER OF $x+2$ IS ALSO 2; SO THEIR LCM IS $(x+1)^2(x+2)^2$. THE SUM BECOMES

$$\frac{(x+1)}{(x+1)^2(x+2)^2} + \frac{(x+2)}{(x+1)^2(x+2)^2}$$

$$= \frac{x+1+x+2}{(x+1)^2(x+2)^2}$$

$$= \frac{2x+3}{(x+1)^2(x+2)^2}$$

AND NOW, AFTER YOU DO SOME PRACTICE PROBLEMS, WE'LL GO ON TO SEE WHAT ALL THIS IS GOOD FOR!

WHAT? MATH IS GOOD FOR SOMETHING?

Problems

1. FIND THE LEAST COMMON MULTIPLE OF:

   a. 4 AND 6   *d.* 72 AND 54

   b. 3 AND 9   *e.* 10 AND 11

   c. 3 AND 7   *f.* 49 AND 21

2. FIND THE LEAST COMMON MULTIPLE OF:

   a. p^2q AND pq^8

   b. x^2 AND x^9

   c. $2a^2x^2(x+1)$ AND $4ax$

   d. x AND x^2+1

   e. $r^5s^3tuv^8$ AND $r^3s^{20}t^9v^4$

   f. $(x-2)^2(x+2)$ AND $(x-2)(x+2)^3(x+3)$

   g. x^2+x+1 AND $x(x^2+x+1)$

   h. $18(x^2+1)^3(x^3-5)^2$ AND $20(x^2+1)^2(x^3-5)^4$

3. MULTIPLY OR DIVIDE

   a. $\dfrac{a}{c}\cdot\dfrac{b}{ad}$

   b. $\dfrac{ax}{c}\cdot\dfrac{bx}{c}$

   c. $\dfrac{x}{b}\div\dfrac{b}{x}$

   d. $\dfrac{\left(\frac{x}{y}\right)}{\left(\frac{1}{y}\right)}$

   e. $\dfrac{3(at)^2}{b}\cdot\dfrac{b^3}{9a}$

   f. $\left(\dfrac{a(x+1)y^{10}}{8pq}\right)\left(\dfrac{2p^3a}{(x+1)^2}\right)$

4. IF $\dfrac{1}{r}+\dfrac{1}{s}=Q$

THEN WHAT IS r IN TERMS OF s AND Q?

5. ADD (OR SUBTRACT):

   a. $\dfrac{a^2}{b^2}+\dfrac{t^2}{b^2}$

   b. $\dfrac{a^3}{2b^2}+\dfrac{5}{b^2}$

   c. $\dfrac{2(x+3)}{(x+1)(x+2)}+\dfrac{x+2}{(x+1)(x+3)}-\dfrac{6(x+1)}{(x+2)(x+3)}$

   d. $\dfrac{x}{b}-\dfrac{b}{x}$

   e. $\dfrac{2}{x}-\dfrac{x}{1+x^2}$

   f. $1+\dfrac{x-1}{x+1}$

   g. $A+\dfrac{B^2-AC}{C}$

   h. $\dfrac{1}{2a+2ax^2}+\dfrac{6}{a^4(1+x^2)^4}$

6. A POSITIVE WHOLE NUMBER IS CALLED **COMPOSITE** IF IT'S THE PRODUCT OF TWO FACTORS SMALLER THAN ITSELF, SUCH AS $12=4\times3$. OTHERWISE, IT'S CALLED **PRIME**. A PRIME NUMBER'S ONLY FACTORS ARE ITSELF AND 1, FOR INSTANCE $3=3\times1$, $17=17\times1$.

IF YOU FACTOR ANY COMPOSITE NUMBER, THEN EACH FACTOR IS EITHER PRIME OR COMPOSITE; THE COMPOSITE ONES CAN BE FURTHER FACTORED... AND SO ON, UNTIL YOU REACH A PRODUCT OF PRIMES ONLY.

$$180 = 10\times18 = (5\times2)\times(6\times3)$$
$$= 5\times2\times(2\times3)\times3 = 2^23^25$$

AS SOME OF THESE PRIME FACTORS MAY APPEAR MORE THAN ONCE, WE SEE THAT ANY NUMBER CAN BE WRITTEN AS THE PRODUCT OF **POWERS OF PRIMES**.

NOW WE CAN DETERMINE THE LCM OF TWO **NUMBERS** IN EXACTLY THE SAME WAY AS WE FOUND THE LCM OF TWO **ALGEBRAIC EXPRESSIONS:** 1. FACTOR EACH NUMBER INTO POWERS OF PRIMES. 2. FIND THE HIGHEST POWER OF EACH PRIME THAT APPEARS. 3. MULTIPLY THOSE POWERS TOGETHER. FOR INSTANCE:

$$36 = 2^23^2 \text{ AND } 24 = 2^33.$$

THE LARGEST EXPONENT OF 2 IS 3; THE LARGEST EXPONENT OF 3 IS 2, SO THE LCM OF 24 AND 36 IS $2^33^2 = 72$.

USING THIS METHOD, FIND THE LCM OF

   a. 36 AND 180   *b.* 225 AND 30

   c. 33 AND 1,617.

7. IF TWO POSITIVE INTEGERS DIFFER BY ONE, CAN THEY HAVE A COMMON MULTIPLE LESS THAN THEIR PRODUCT?

Chapter 11
Rates

THINGS ARE ALWAYS GETTING BETTER OR WORSE OR
BIGGER OR SMALLER. THE QUESTION IS, HOW FAST?

JUST TO PROVE THAT ALGEBRA CAN BE A PIECE OF CAKE, HERE'S ONE NOW. THIS PARTICULAR
SLICE HAPPENS TO PLAY HOST TO A SMALL, HUNGRY INSECT: A REMARKABLE CAKE WEEVIL THAT
MUNCHES STEADILY ON, NEVER SPEEDING UP OR SLOWING DOWN, FOR AS LONG AS THERE'S
CAKE TO BE EATEN. (SOMEHOW, THIS HUNGRY BUG NEVER GETS FULL.)

IN EVERY MINUTE, THE WEEVIL EATS EXACTLY 2 OUNCES (OZ.) OF CAKE. IN 2 MINUTES, IT EATS TWICE THAT, OR 4 OZ.; IN 3 MINUTES, IT EATS $2 \times 3 = 6$ OZ.; AND SO FORTH, AS SHOWN IN THIS TABLE:

TIME ELAPSED (MINUTES)	CAKE EATEN (OUNCES)
1	2
2	4
3	6
4	8
5	10
6	12
ETC.	

THE **RATE** AT WHICH THE WEEVIL EATS CAKE IS A QUOTIENT: THE AMOUNT OF CAKE EATEN (MEASURED IN OUNCES) IN A GIVEN TIME DIVIDED BY THE LENGTH OF TIME.

$$\text{RATE OF EATING} = \frac{\text{AMOUNT EATEN}}{\text{ELAPSED TIME}}$$

IF WE DO THE DIVISION ALONG ANY LINE OF THE TABLE, WE ALWAYS GET THE SAME ANSWER: 2. WE SAY THAT THE WEEVIL'S RATE IS...

OR **2 OUNCES PER MINUTE.** THE SLASH MARK / INDICATES THAT THE RATE COMES FROM DIVISION.

BY NOW WE RECOGNIZE THE PHRASES "AMOUNT OF CAKE," "ELAPSED TIME," AND "RATE OF EATING" AS WHAT THEY ARE: **VARIABLES.** THIS BEING AN ALGEBRA BOOK, WE USE A SINGLE LETTER FOR EACH.

t = ELAPSED TIME
E = AMOUNT OF CAKE EATEN IN TIME t
r = RATE

THEN THE EQUATION DEFINING THE RATE LOOKS LIKE THIS:

$$r = \frac{E}{t}$$

MULTIPLYING BOTH SIDES BY t PUTS IT IN THIS FORM:

$$E = rt$$

THE AMOUNT EATEN, E, IS THE PRODUCT OF THE RATE AND THE TIME—EVEN WHEN t IS NOT AN INTEGER. IN HALF A MINUTE ($t = \frac{1}{2}$), AT A RATE OF 2 OZ/MIN, THE WEEVIL EATS $(2) \cdot (\frac{1}{2}) = 1$ OUNCE. IN 7.16 MINUTES, IT WOULD BE $(2)(7.16) = 14.32$ OUNCES. IF THE WEEVIL ATE FASTER, SAY AT A RATE OF 2.4 OUNCES PER MINUTE, THEN IN 6 MINUTES IT WOULD EAT $(2.4)(6) = 14.4$ OUNCES, AND SO ON. IT'S AUTOMATIC!

I'M NOT A VERY SPONTANEOUS GRUB...

OW!

AND AT 2 OZ/MIN, HOW MANY OUNCES GO DOWN IN 35 **SECONDS?** HERE WE HAVE TO DO A LITTLE ARITHMETIC TO CONVERT SECONDS TO MINUTES.

$$35 \text{ SEC.} = \frac{35}{60} \text{ MIN.}$$

SO THE AMOUNT EATEN WOULD BE

$$2 \cdot \left(\frac{35}{60}\right) = \frac{70}{60} = \frac{7}{6} \text{ OZ.}$$

RATES ARE EVERYWHERE IN THE WORLD, NOT ONLY IN PLACES WEEVILS GO. FOR EXAMPLE,

Wages: JESSE BABYSITS FOR A PAY RATE OF $8.75 PER HOUR. HE GETS $8.75 × THE NUMBER OF HOURS WORKED.

MY BILL FOR 3.162 HOURS, MR. AND MRS. DUDE...

Amount due: $27.6675

Fluid Flow: AS WATER POURS INTO A BATHTUB, THE **RATE OF FLOW** IS THE VOLUME OF WATER ADDED PER UNIT TIME (IN GALLONS PER MINUTE, SAY).

SOON WE'LL GO OVER THE TOP AND BUST OUT!

Speed: A CAR TRAVELS SOME NUMBER OF MILES EVERY HOUR. THE RATE IS ITS **SPEED**, DISTANCE TRAVELED DIVIDED BY ELAPSED TIME:

$$\text{SPEED} = \frac{\text{DISTANCE}}{\text{TIME}}$$

A SPEEDO-METER KNOWS HOW TO DIVIDE?

Price: WHEN YOU BUY GASOLINE, YOU PAY AT A COST PER GALLON. THE POSTED PRICE PER GALLON IS REALLY A RATE.

$$\text{PRICE PER GALLON} = \frac{\text{TOTAL COST}}{\text{VOLUME OF GAS}}$$

A RATE MEASURED AGAINST SOMETHING OTHER THAN TIME!

3.79
3.84

Sports: IN BASEBALL, A PLAYER'S **BATTING AVERAGE** IS THE NUMBER OF HITS MADE DIVIDED BY THE NUMBER OF AT-BATS. IT'S THE RATE OF HITS PER AT BAT.

$$\text{BATTING AVERAGE} = \frac{\text{HITS}}{\text{AT-BATS}}$$

AND THE RATE OF STEROID USE PER PLAYER-POUND?

LET'S RETURN TO THE CAKE-EATING EQUATION $E = rt$. WE CAN DRAW A GRAPH OF THIS EQUATION.

t	E
1	r
2	2r
3	3r
4	4r
5	5r
6	6r

ETC.

$E = rt$

THE GRAPH IS A STRAIGHT LINE, AND r IS ITS **SLOPE**. THE SLOPE ITSELF IS A RATE. IT'S THE RATE AT WHICH A LINE RISES OR FALLS PER HORIZONTAL UNIT.

RISE OVER RUN, REMEMBER?

THIS RAISES A QUESTION: CAN A RATE EQUATION'S GRAPH EVER SLOPE DOWNWARD? IS THERE SUCH A THING AS A **NEGATIVE RATE?**

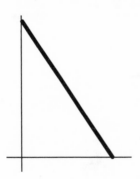

ANSWER: YES. THE RATE IS NEGATIVE WHEN SOMETHING **DECREASES**. FOR INSTANCE, AS WATER DRAINS **OUT** OF A TUB, THE AMOUNT OF WATER IN THE TUB GOES DOWN, AND ITS RATE OF CHANGE IS NEGATIVE.

GRGGL

IN THE SAME WAY, WHEN A WEEVIL EATS CAKE AT **2** OZ/MIN., THE AMOUNT OF **UNEATEN CAKE** CHANGES AT A RATE OF **−2** OZ/MIN.

MOMENT BY MOMENT, THERE'S LESS AND LESS...

THEN WHAT'S THE EQUATION FOR THE UNEATEN CAKE (CALL IT U)? NOT $U = rt$, BECAUSE THAT WOULD MAKE NEGATIVE CAKE AFTER POSITIVE TIME, AND POSITIVE CAKE IS STILL THERE...

BURP!

139

The ALL-PURPOSE Rate Equation

HOW CAN WE FIND AN EQUATION FOR THE RATE CHANGE OF UNEATEN CAKE? START WITH WHAT WE KNOW: THE TOTAL AMOUNT OF CAKE IS THE SUM OF THE UNEATEN CAKE U AND THE CAKE INSIDE THE WEEVIL, E.

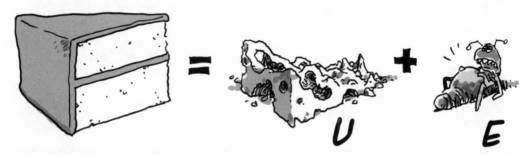

NOTE THAT THERE'S NO SYMBOL YET FOR THE TOTAL AMOUNT OF CAKE. WE'RE GOING TO CALL IT SOMETHING RATHER STRANGE-LOOKING:

("YOU-NOUGHT"). THIS INDICATES THAT IT WAS THE AMOUNT OF UNEATEN CAKE AT THE **BEGINNING**, AT "TIME ZERO," BEFORE THE WEEVIL STARTED TO EAT.

TIME ZERO

THE EQUATION ABOVE BECOMES

$$U_0 = U + E$$

$$E = U_0 - U$$

OR

BECAUSE WE WILL BE LOOKING AT DIFFERENT RATES, LET'S WRITE r_E, INSTEAD OF JUST r, FOR THE RATE OF EATING. THE BASIC RATE EQUATION ON P. 137 NOW LOOKS LIKE THIS:

$$r_E = \frac{U_0 - U}{\text{ELAPSED TIME}}$$

AND WHAT IS ELAPSED TIME? YOU CAN'T READ IT ON A CLOCK... INSTEAD YOU HAVE TO TAKE THE **DIFFERENCE** BETWEEN THE TIME t **NOW** AND THE **INITIAL TIME** t_0 WHEN THE WEEVIL BEGAN TO EAT AND THE AMOUNT OF CAKE WAS U_0.

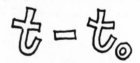

SO THE ELAPSED TIME IS

$$t - t_0$$

AND THE WEEVIL'S RATE OF CAKE-EATING IS

$$r_E = \frac{U_0 - U}{t - t_0}$$

FINALLY, WE KNOW ONE MORE THING: r_U, THE RATE OF CHANGE OF **UNEATEN** CAKE, IS THE **NEGATIVE** OF r_E! IT MUST BE SO: WHATEVER GOES INTO THE WEEVIL IN A GIVEN TIME COMES OFF THE CAKE IN THE SAME TIME!

$$r_U = -r_E$$

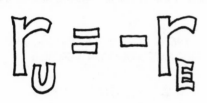

NOW DO THE ALGEBRA:

$$r_U = -r_E = -\left(\frac{U_0 - U}{t - t_0}\right)$$

$$= \frac{-(U_0 - U)}{t - t_0} = \frac{U - U_0}{t - t_0}$$

SO HERE IT IS:

$$r_U = \frac{U - U_0}{t - t_0}$$

r_U IS THE **CHANGE IN U** FROM TIME t_0 TO TIME t DIVIDED BY THE **CHANGE IN TIME.**

MULTIPLYING THROUGH BY THE QUANTITY $(t - t_0)$ GIVES

$$U = U_0 + r_U(t - t_0)$$

THIS IS THE **ALL-PURPOSE RATE EQUATION:** IT SAYS THAT THE AMOUNT OF STUFF AT TIME t EQUALS THE ORIGINAL AMOUNT OF STUFF PLUS THE RATE TIMES THE TIME CHANGE.

Example 1.

FIND THE RATE EQUATION FOR UNEATEN CAKE U WHEN $U_0 = 80$ OZ, $r_U = -3$ OZ/MIN, $t_0 =$ MIDNIGHT.

LET'S CALL MIDNIGHT "ZERO O'CLOCK," SO $t_0 = 0$. THE ALL-PURPOSE EQUATION IS

$$U = U_0 + r_U(t - t_0)$$

PLUGGING IN THE GIVEN VALUES GIVES

$$U = 80 + (-3)(t - 0)$$

$$U = 80 - 3t$$

t	U
5	65
10	50
15	35
20	20
25	5

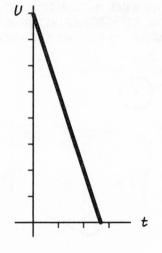

FROM THIS WE CAN MAKE A TABLE OF VALUES OF U AT DIFFERENT TIMES t (MEASURING t IN MINUTES AFTER MIDNIGHT) AND A GRAPH OF THE EQUATION.

FOR INSTANCE, AT 25 MINUTES AFTER MIDNIGHT, ONLY 5 OUNCES OF CAKE REMAIN UNEATEN.

Example 2.

THE ALL-PURPOSE RATE EQUATION ALSO APPLIES TO E, THE AMOUNT EATEN BY THE WEEVIL. IT SAYS:

$$E = E_0 + r_E(t - t_0)$$

E_0 IS THE AMOUNT OF CAKE ALREADY EATEN BY THE WEEVIL AT TIME t_0 (FROM AN EARLIER PIECE OF CAKE, SAY).

SUPPOSE THAT $E_0 = 2$ OZ., AND THE WEEVIL EATS AT A STEADY RATE OF 1.6 OZ/MIN. IF t_0 IS 12:30 PM, THEN HOW MUCH CAKE IS IN THE WEEVIL AT TIME t?

WITH THOSE VALUES FOR t_0, E_0, AND r_E, THE RATE EQUATION BECOMES

$$E = 2 + (1.6)(t - 12{:}30)$$

AGAIN WE CAN DRAW ITS GRAPH, WHICH SHOWS HOW MUCH CAKE THE WEEVIL HAS EATEN AT DIFFERENT TIMES.

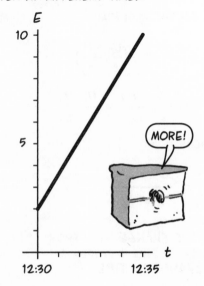

THE ALL-PURPOSE RATE EQUATION HAS AN
ALL-PURPOSE GRAPH. SUPPOSE A IS THE
AMOUNT OF SOMETHING CHANGING AT RATE
r, AND t IS TIME. (t CAN ACTUALLY BE ANY
VARIABLE ON WHICH A DEPENDS.) THE INITIAL
VALUES OF t AND A ARE t_0 AND A_0. THE
ALL-PURPOSE RATE EQUATION SAYS:

$$A = A_0 + r(t - t_0) \quad \text{OR}$$

$$A - A_0 = r(t - t_0)$$

THIS MAY LOOK FAMILIAR. IT IS THE
**POINT-SLOPE FORM OF THE EQUATION
OF A LINE PASSING THROUGH THE
POINT (t_0, A_0) WITH SLOPE r.**

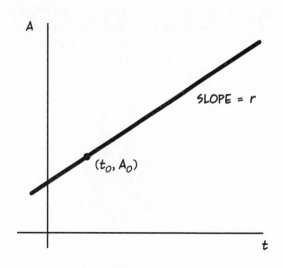

SLOPE = r

(t_0, A_0)

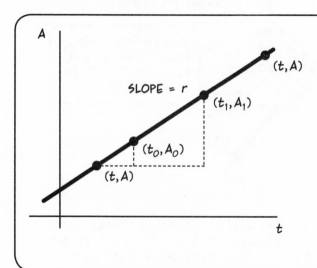

SLOPE = r

(t, A)

(t_1, A_1)

(t_0, A_0)

(t, A)

THIS TELLS US THAT IF (t_1, A_1) IS
ANY POINT ON THE GRAPH, THEN
THE EQUATION IS STILL TRUE USING
(t_1, A_1) IN PLACE OF (t_0, A_0).

$$A = A_1 + r(t - t_1)$$

IN OTHER WORDS, THE ALL-PURPOSE
RATE EQUATION IS GOOD **NO MATTER
WHAT TIME WE CHOOSE FOR THE
STARTING TIME.** NOT ONLY THAT,
BUT THE EQUATION IS ALSO TRUE
WHETHER $t < t_1$ OR $t > t_1$.

THE SLOPE r EQUALS THE
RISE OVER THE RUN, NO
MATTER WHICH TWO POINTS
ON THE LINE ARE USED!

CAKE
HOG...

SPEED & VELOCITY

SPEED, WE SAID, IS A RATE: IT'S DISTANCE DIVIDED BY TIME. **SPEED IS ALWAYS A POSITIVE NUMBER.**

AND THIS IS A PROBLEM... BECAUSE IN MATH WE ALMOST ALWAYS WANT RATES THAT CAN BE POSITIVE **OR** NEGATIVE.

I'VE NEVER SEEN A SPEEDOMETER WITH ANYTHING ELSE...

SPROING

SIGH... NATURALLY!

JUST AS AN AMOUNT OF CAKE CAN INCREASE OR DECREASE, THE RATE OF MOTION SHOULD SAY WHETHER A MOVING OBJECT IS GOING UP OR DOWN, FORWARD OR BACKWARD.

THERE'S **SOMETHING** NEGATIVE ABOUT THIS, BUT I CAN'T QUITE PUT MY FINGER ON IT...

IMAGINE A STRAIGHT ROAD STRETCHING ENDLESSLY IN BOTH DIRECTIONS (A NUMBER LINE!). TAKE SOME POINT ON THE ROAD TO BE s_0, THE STARTING POINT. A STEADILY MOVING CAR PASSES THROUGH s_0 AT TIME t_0. SUPPOSE t IS ANY OTHER TIME, AND s IS THE CAR'S POSITION AT TIME t.

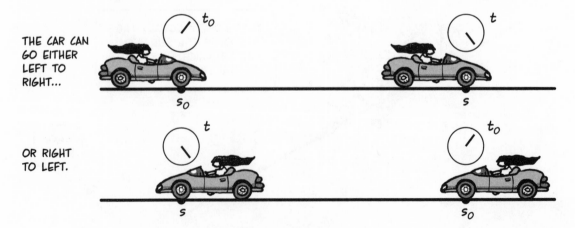

THE CAR CAN GO EITHER LEFT TO RIGHT...

OR RIGHT TO LEFT.

INSTEAD OF DISTANCE, WE THINK IN TERMS OF **CHANGE OF POSITION,** $s - s_0$.* WHEN MOVING FORWARD, $s - s_0 > 0$, AND IT'S THE SAME AS DISTANCE. WHEN MOVING BACKWARD, $s - s_0 < 0$, THE NEGATIVE OF DISTANCE.

DISTANCE = $|s - s_0|$

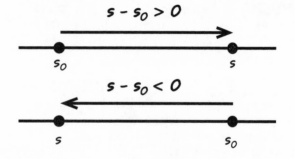

$$s - s_0 > 0$$

$$s - s_0 < 0$$

*THE LETTER s STANDS FOR *SITUS*, LATIN FOR *PLACE*. ONCE UPON A TIME, ALL EDUCATED PEOPLE LEARNED LATIN, WHICH LET THEM COMMUNICATE ACROSS NATIONAL BORDERS AND ACROSS TIME, TOO. HARDLY ANYONE STUDIES LATIN ANYMORE, BUT LATIN INITIALS STILL HAUNT US LIKE GHOSTS...

THE CAR'S **VELOCITY v** IS THE RATE OF **CHANGE OF POSITION** PER UNIT TIME.

$$V = \frac{s - s_0}{t - t_0}$$

VELOCITY IS **EQUAL** TO SPEED WHEN MOVING FORWARD; IT'S THE **NEGATIVE** OF SPEED WHEN MOVING BACKWARD. PEOPLE OFTEN DESCRIBE VELOCITY AS **"SPEED WITH DIRECTION."**

$t_0 < t, \ s_0 < s, \ v > 0$

$t_0 < t, \ s_0 > s, \ v < 0$

$t_0 > t, \ s_0 < s, \ v < 0$

$t_0 > t, \ s_0 > s, \ v > 0$

THE ALL-PURPOSE VELOCITY EQUATION DESCRIBES POSITION IN TERMS OF VELOCITY AND TIME.

$$s = s_0 + v(t - t_0)$$

Example 3.

STARTING 30 MILES TO CELIA'S EAST, A CAR TRAVELS TO A POINT 100 MILES TO HER WEST. THE TRIP TAKES 2 HOURS. WHAT WAS THE CAR'S VELOCITY?

WE TAKE CELIA'S POSITION TO BE ZERO; EAST (RIGHT) IS POSITIVE AND WEST (LEFT) IS NEGATIVE.

PURE PREJUDICE!

THE GIVENS: $s_0 = 30$, $s = -100$, $t - t_0 = 2$ HRS. (REMEMBER $t - t_0$ IS ALWAYS ELAPSED TIME!) THEN

$$v = \frac{-100 - 30}{2} = -65 \text{ MI/HR}$$

NEGATIVE VELOCITY MEANS WESTWARD MOVEMENT.

SOUTHIST!

Example 4.

CELIA WALKS EASTWARD AT A SPEED OF 1.4 MI/HR. IF SHE STARTS 2 MILES WEST OF ME AT 2:30 PM, WHERE WILL SHE BE AT 5:45?

THE GIVENS: $t_0 = 2:30$, $t = 5:45$, $s_0 = -2$, $v = 1.4$, WITH MY POSITION AT $s = 0$.

ANSWER: FIRST FIND $t - t_0$, THE ELAPSED TIME. IT IS

$$5:45 - 2:30 = 3\tfrac{1}{4} \text{ HRS} = \frac{13}{4} \text{ HRS.}$$

THE ALL-PURPOSE RATE EQUATION SAYS

$$s = s_0 + v(t - t_0) = -2 + (1.4)(\tfrac{13}{4})$$

$$= -2 + 4.55 = 2.55$$

AT 5:45 PM, CELIA WILL BE **2.55** MILES **EAST** OF ME (AS INDICATED BY THE POSITIVE SIGN OF THE ANSWER).

MUSTN'T SPEED UP OR SLOW DOWN... MUST KEEP WALKING... WALKING...

Example 5. TWO BANK ROBBERS MAKE THEIR GETAWAY AT HIGH NOON, MOTORING EAST AT 70 MI/HR. THE POLICE, BUSY WITH LUNCH, GET MOVING AT 1 PM. IF THE POLICE STATION IS 6 MILES WEST OF THE BANK, AND THE COPS DRIVE 90 MI/HR., WHEN AND WHERE DO THEY OVERTAKE THE CROOKS? TAKE THE BANK'S POSITION TO BE 0, AND LET $t_0 = 12$ NOON.

NEVER CHASE ON AN EMPTY STOMACH...

POLICE
Mile -6 Station

O

VROooo.

WE BEGIN BY WRITING SEPARATE RATE EQUATIONS FOR THE POLICE AND THE CROOKS. LETTING s_C BE THE CROOKS' POSITION, THEIR EQUATION IS

$$s_C = 0 + 70(t - t_0)$$

$$= 70(t - t_0)$$

THE POLICE BEGIN MOVING AN HOUR LATER, AT TIME $t_0 + 1$ HR. THEIR INITIAL POSITION IS -6, SO THEIR POSITION s_P AT TIME t IS

$$s_P = -6 + 90(t - (t_0 + 1))$$

$$= 90(t - t_0) - 96$$

THE COPS CATCH THE CROOKS WHEN THEY HAVE **THE SAME POSITION**, THAT IS, WHEN $s_C = s_P$.

SO SET $s_C = s_P$ AND SOLVE FOR t.

$$70(t - t_0) = 90(t - t_0) - 96$$

$$20(t - t_0) = 96$$

$$t - t_0 = \frac{96}{20} = \textbf{4.8 HOURS}$$

SINCE t_0 IS THE **CROOKS'** STARTING TIME, THIS EQUATION SAYS THAT THEY ARE CAUGHT AT NOON PLUS 4.8 HOURS, OR **4:48** PM. (0.8 HOURS = $60 \times 0.8 = 48$ MINUTES.)

AND **WHERE** ARE THEY CAUGHT? EITHER EQUATION WILL TELL US. THE CROOKS' EQUATION IS EASIER:

$$s_C = (70)(4.8) = \textbf{336}$$

THEY ARE CAUGHT AT MILE 336, THAT IS, 336 MILES EAST OF THE BANK (AND $342 = 336 + 6$ MILES EAST OF THE POLICE STATION).

HELLO? JIFFY-TOW?

I HATE IT WHEN THAT HAPPENS...

COMBINING Rates

IN THE LAST PROBLEM, TWO
DIFFERENT RATES CAME INTO
PLAY. IS THERE A WAY TO
COMBINE RATES?

SUPPOSE TWO WEEVILS ARE BOTH EATING THE SAME PIECE OF CAKE. IF THE SLOWER
WEEVIL EATS **2** OZ/MIN AND THE FASTER ONE'S RATE IS **3** OZ/MIN, THEN IT'S PRETTY
CLEAR THAT **5** OUNCES OF CAKE ARE EATEN EVERY MINUTE.

IN GENERAL, IF WEEVIL 1 EATS AT THE RATE r_1, AND WEEVIL 2 EATS AT THE RATE r_2, THEN
THE COMBINED RATE **r** IS THE **SUM:**

$$r = r_1 + r_2$$

Example 6.

SUPPOSE WATER FLOWS INTO A 500-LITER TANK AT 2 LITERS PER MINUTE (L/MIN). AT THE SAME TIME, THE TANK LEAKS WATER AT A RATE OF $\frac{1}{3}$ L/MIN. IF THE TANK CONTAINS 100 LITERS RIGHT NOW, HOW LONG WILL IT TAKE TO FILL UP?

WE HAVE TWO RATES HERE, A RATE r_1 OF WATER FLOWING IN, AND A RATE r_2 OF WATER FLOWING OUT.

$$r_1 = 2 \text{ L/MIN}, \qquad r_2 = -\frac{1}{3} \text{ L/MIN}$$

(r_2 IS NEGATIVE BECAUSE WATER FLOWING OUT DECREASES THE VOLUME.)

THE COMBINED RATE r IS THEIR SUM.

$$r = 2 - \frac{1}{3} = \frac{5}{3} \text{ L/MIN}$$

IF V IS THE VOLUME AT TIME t, $V_0 = 100$ L THE ORIGINAL VOLUME, AND $t_0 =$ NOW OR 0, THEN THE RATE EQUATION $V = V_0 + rt$ GIVES

$$V = 100 + \frac{5}{3} t$$

WE WANT TO KNOW THE TIME t WHEN V IS 500, SO SET $V = 500$ AND SOLVE FOR t.

$$500 = 100 + \frac{5}{3} t$$

$$\frac{5}{3} t = 400$$

$$t = \frac{1200}{5}$$

$$= \mathbf{240} \text{ MIN, OR 4 HOURS.}$$

ANOTHER WAY to Describe Rates

SOMETIMES RATES COME TO US "UPSIDE DOWN." WHEN MOMO TELLS ME SHE CAN MOW THIS LAWN IN SIX HOURS, WHAT IS HER RATE? WE FIND IT BY DIVIDING THE AMOUNT OF MOWED LAWN BY THE LENGTH OF TIME IT TAKES.

$$\text{RATE} = \frac{\text{AMOUNT OF LAWN}}{\text{TIME}}$$

$$\text{RATE} = \frac{1 \text{ LAWN}}{6 \text{ HOURS}} = \frac{1}{6} \text{ LAWN/HOUR}$$

THIS IS A PERFECTLY GOOD WAY TO DESCRIBE A RATE! ALGEBRAICALLY, IF WE'RE GIVEN THE TIME T TAKEN TO DO ONE JOB, THEN WE INVERT THE TIME TO FIND THE RATE IN TERMS OF JOBS PER UNIT OF TIME.

$$1 \text{ JOB} = rT$$

$$r = \frac{1 \text{ JOB}}{T \text{ TIME UNITS}}$$

(1) $\quad r = \dfrac{1}{T} \text{ JOB/TIME UNIT}$

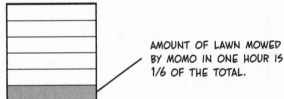

AMOUNT OF LAWN MOWED BY MOMO IN ONE HOUR IS 1/6 OF THE TOTAL.

Example 7.
NOW KEVIN BRINGS A BIG, HONKING POWER MOWER AND OFFERS TO HELP MOMO. WORKING ALONE, HE CAN DO THE JOB IN JUST 2 HOURS. HOW LONG DOES IT TAKE IF THEY WORK TOGETHER?

STARTING FROM OPPOSITE ENDS, OF COURSE!

HONK

WHAT?

SOLUTION: LET r_M BE MOMO'S RATE AND r_K BE KEVIN'S RATE. THEN THE COMBINED RATE r IS THEIR SUM:

$$r = r_M + r_K$$

WE ARE GIVEN THIS:

$$r_M = \frac{1}{6} \qquad r_K = \frac{1}{2}$$

THE SUM IS

$$\frac{1}{6} + \frac{1}{2} = \mathbf{\frac{2}{3}} \text{ LAWNS/HR}$$

AGAIN LETTING T BE THE AMOUNT OF TIME IT TAKES TO DO ONE JOB, WE GO BACK TO EQUATION 1:

$$r = 1/T$$

MULTIPLYING BY T/r GIVES

$$T = 1/r$$

THAT IS, THE TIME IS THE RECIPROCAL OF THE RATE—SO THE JOB TAKES

$$(2/3)^{-1} = \mathbf{\frac{3}{2}} \text{ HOURS}$$

I.E., AN HOUR AND A HALF.

HONK
HONK
HONK

WE CAN ALSO ASK WHAT PART OF THE LAWN EACH OF THEM MOWED. WE FIND THIS BY MULTIPLYING EACH INDIVIDUAL'S RATE BY THE TIME WORKED, THAT IS, 3/2 HOURS.

MOMO MOWS $\frac{1}{6} \cdot \frac{3}{2} = \frac{1}{4}$ LAWN

KEVIN MOWS $\frac{1}{2} \cdot \frac{3}{2} = \frac{3}{4}$ LAWN

KEVIN MOWS THREE TIMES AS MUCH AREA AS MOMO, WHICH IS NOT SURPRISING, CONSIDERING THAT HIS RATE IS THREE TIMES HERS.

THIS EXAMPLE HAS BEEN BROUGHT TO YOU BY THE HONKMASTER CORPORATION...

A Sense of PROPORTION

THE VERY SIMPLEST RATE EQUATION
RELATING TWO VARIABLES x AND y IS

$$y = Cx$$

WHERE C IS SOME CONSTANT, SUCH AS
1, 2, OR 150. IN THIS EQUATION, WE
SAY THAT y IS **PROPORTIONAL** TO x.
WHEN THIS IS TRUE, AND (x_1, y_1) AND
(x_2, y_2) ARE ANY PAIRS OF VALUES
SATISFYING THE EQUATION, THEN

$$\frac{y_1}{x_1} = \frac{y_2}{x_2} = C$$

C IS CALLED THE **PROPORTIONALITY
CONSTANT.**

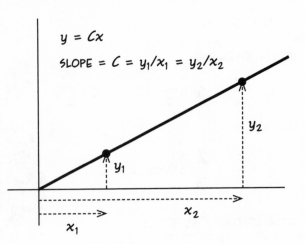

$y = Cx$

$$\text{SLOPE} = C = y_1/x_1 = y_2/x_2$$

Example 8. WHEN RESIZING AN IMAGE, AN ENLARGEMENT OR REDUCTION IS PROPOR-
TIONAL WHEN THE RATIO BETWEEN HEIGHT AND WIDTH IS MAINTAINED: THEY ARE SCALED BY
THE SAME FACTOR. ENLARGING BY 200%, FOR INSTANCE, MEANS DOUBLING BOTH HEIGHT AND
WIDTH. A **DIS**PROPORTIONATE CHANGE WOULD SCALE HEIGHT AND WIDTH DIFFERENTLY.

PROPORTIONAL:

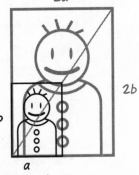

NOT PROPORTIONAL:

IN PROPORTIONAL SCALING, THE LENGTH OF ANY FEATURE IN THE IMAGE IS SCALED BY THE
SAME FACTOR—DOUBLED IN OUR PROPORTIONAL PICTURE. WHEN SCALING IS DISPROPOR-
TIONATE, DIFFERENT FEATURES SCALE DIFFERENTLY.

IN THAT PICTURE,
WIDTHS GROW
MORE THAN
HEIGHTS, CIRCLES
FLATTEN OUT...

DO
TELL...

Example 9. HERE IS ANOTHER USE OF PROPORTION. SUPPOSE WE KNOW KEVIN'S HEIGHT, THE LENGTH OF HIS SHADOW, AND THE LENGTH OF A TREE'S SHADOW. THEN WE CAN FIND THE TREE'S HEIGHT.

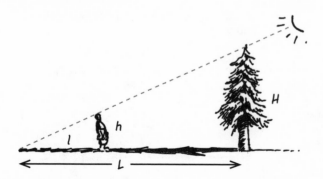

TO SEE THE PROPORTIONALITY, KEVIN STANDS SO THAT HIS HEAD, THE TREETOP, AND THE SUN ALL LINE UP. THEN THE RATIO OF HEIGHT TO SHADOW LENGTH IS THE SLOPE OF THIS LINE, WHETHER WE'RE TALKING ABOUT KEVIN, THE TREE, A SMALL UPRIGHT STICK, OR ANYTHING.

FOR THE ALGEBRA, LET

h = KEVIN'S HEIGHT
H = TREE'S HEIGHT
l = KEVIN'S SHADOW'S LENGTH
L = TREE'S SHADOW LENGTH

THEN

$$\frac{H}{L} = \frac{h}{l}$$

MULTIPLYING BOTH SIDES BY L, WE GET

$$H = L\frac{h}{l}$$

IF, FOR INSTANCE, KEVIN IS 1.8 METERS TALL, HIS SHADOW IS 2.5 METERS LONG, AND THE TREE'S SHADOW IS 34 METERS LONG, THEN

$$\frac{H}{34} = \frac{1.8}{2.5}, \quad H = \frac{(1.8)(34)}{2.5}$$

$H = 24.48$ METERS IS THE TREE'S HEIGHT.

LIGHT HAS BEEN SHED!

Important thing to remember, which you may already know but which always bears repeating:

IF A, a, B, AND b ARE IN PROPORTION, IN OTHER WORDS, IF $B/A = b/a$, (AND a, b, A, AND B ARE NOT ZERO), THEN ALSO

$$Ab = aB, \quad \frac{A}{a} = \frac{B}{b}, \quad \frac{a}{A} = \frac{b}{B}, \quad \frac{a}{b} = \frac{A}{B}$$

IF YOU KNOW ANY THREE VALUES, YOU CAN FIND THE FOURTH.

Problems

1. MOMO BABYSITS FOR $3\frac{1}{2}$ HOURS AND IS PAID $19.25. WHAT IS HER HOURLY PAY RATE?

2. I FILL MY CAR WITH GASOLINE AT $3.69/GALLON. THE COST IS $44.28. IF MY GAS TANK'S TOTAL CAPACITY IS 15 GALLONS, HOW MUCH GAS WAS IN THE TANK WHEN I STARTED TO PUMP?

3. IF A PIECE OF CAKE WEIGHS 14 OUNCES, AND IT TAKES A WEEVIL 6 MINUTES TO EAT IT, WHAT IS THE WEEVIL'S RATE OF EATING, IN OUNCES PER MINUTE? WHAT IS THE RATE IN PIECES PER MINUTE?

4. A CAKE WEIGHS 500 GRAMS. A WEEVIL BEGINS EATING AT A RATE OF 15 G/MIN AT 6:45 AM. HOW MUCH CAKE IS LEFT AT 7:10?

5. A WEEVIL IS EATING A PIECE OF CAKE. IF THERE ARE 3 OUNCES OF CAKE NOW, AND THE WEEVIL WAS EATING AT A RATE OF 2 OZ/MIN, HOW MUCH CAKE WAS THERE 10 MINUTES AGO?

6a. CELIA CAN MOW A CERTAIN LAWN IN 3 HOURS. JESSE CAN MOW THE SAME LAWN IN 2 HOURS. HOW LONG DOES IT TAKE THEM TO MOW THIS LAWN WORKING TOGETHER? HOW LONG WOULD IT TAKE THEM TO MOW A LAWN TWICE AS BIG?

6b. HOW LONG WOULD IT TAKE IF JESSE STARTS MOWING HALF AN HOUR AFTER CELIA?

7. JESSE CAN CAN MOW A LAWN IN A TIME PERIOD p. MOMO CAN MOW THE SAME LAWN IN A TIME PERIOD q. HOW LONG, IN TERMS OF p AND q, DOES IT TAKE THEM TO MOW THE LAWN TOGETHER?

8. TWO CARS ARE 120 MILES APART. THEY BEGIN DRIVING TOWARD EACH OTHER AT THE SAME TIME. ONE CAR GOES 70 MI/HR; THE OTHER GOES 80 MI/HR.

a. USE THE RATE EQUATION TO FIND WHEN AND WHERE THEY MEET.

b. THINK OF THE CARS AS "EATING UP THE ROAD" BETWEEN THEM. IS THERE ANOTHER WAY TO SOLVE THIS PROBLEM?

9. IT TAKES JESSE 30 SECONDS TO RUN FROM POINT A TO POINT B. IT TAKES CELIA 25 SECONDS TO RUN THE SAME DISTANCE. IF HE BEGINS AT A AND SHE BEGINS AT B, WHEN AND WHERE DO THEY MEET IF THEY BEGAN RUNNING AT THE SAME TIME? WHEN AND WHERE DO THEY MEET IF SHE STARTS 5 SECONDS AFTER HE DOES?

10. MOMO IS 54 INCHES TALL. HER SHADOW IS 27 INCHES LONG. AT THE SAME TIME, SHE MEASURES A TREE'S SHADOW TO BE 41 FEET LONG. HOW TALL IS THE TREE?

11. CELIA, STANDING ON THE BEACH, SEES A SHIP IN THE WATER. NEARBY SHE ALSO SEES A BUOY KNOWN TO BE 100 YARDS OFFSHORE. IS THERE A WAY SHE CAN FIND HOW FAR AWAY THE SHIP IS?

12. A RECTANGLE OF WIDTH a AND HEIGHT b HAS ITS LOWER LEFT CORNER AT THE ORIGIN. WHAT IS THE EQUATION OF THE DIAGONAL LINE RUNNING FROM LOWER LEFT TO UPPER RIGHT?

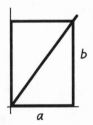

13. IT TAKES ONE PERSON 20 MINUTES TO DIG A HOLE IN THE GROUND. COULD 20 PEOPLE REALLY DIG THE SAME HOLE IN ONE MINUTE?

Chapter 12
About Average

I WROTE THIS CHAPTER
BECAUSE OF A FRUSTRATING
EXPERIENCE I ONCE HAD.
IT WAS, YOU MIGHT SAY, A
WORSE-THAN-AVERAGE
EXPERIENCE, AND I'D LIKE
TO SAVE YOU FROM EVER
HAVING TO SUFFER
THROUGH IT YOURSELF.

THE PROBLEM BEGAN WITH
AN **ELECTRIC BILL,** AND
IT SHOWS HOW GOOD IT
CAN BE TO HAVE SOME-
THING TO PLUG INTO
BESIDES A WALL OUTLET.

IN A BUILDING WHERE I USED TO HAVE STUDIO SPACE, THE **ELECTRIC BILL** WAS DIVIDED AMONG SEVERAL DIFFERENT TENANTS. THE SPLIT DEPENDED (MORE OR LESS) ON ACTUAL USAGE, SO WE ALL PAID A DIFFERENT FRACTION OF THE BILL EVERY MONTH.

STOP USING THE ELECTRIC PENCIL SHARPENER, GONICK!

AFTER SOME TIME, ONE OF THE TENANTS—CALL HIM P*****—ASKED TO MEET WITH THE REST OF US TO DISCUSS THE ELECTRICITY. HIS RECENT BILLS LOOKED SOMETHING LIKE THIS.

HE PAID **14%** IN JUNE;

" " **17%** IN JULY;

" " **14%** IN AUGUST;

" " **25%** IN SEPTEMBER;

" " **26%** IN OCTOBER;

" " **30%** IN NOVEMBER;

" " **28%** IN DECEMBER.

"I TAKE THE AVERAGE," HE SAID—MEANING THAT HE ADDED UP THE SEVEN NUMBERS AND DIVIDED BY 7—

$$\frac{14+17+14+25+26+30+28}{7}$$

"AND I GET

22%"

WHY DON'T I JUST PAY 22% EVERY MONTH?

SEVERAL OF THE OTHER TENANTS SAW THAT THERE WAS AN **ERROR** IN P*****'S REASONING, BUT WHEN WE TRIED TO EXPLAIN IT TO HIM, HE REFUSED TO LISTEN. IT GOT VERY HEATED.

I FIGURED IT OUT **MATHEMATICALLY!**

OF COURSE, THE QUESTION IS: WHAT WAS P*****'S MISTAKE?

NOTHING.

GRRR...

GRR...

GRRRR...

Heights

WE ALL HAVE AN IDEA ABOUT WHAT IT MEANS TO BE AVERAGE. AN AVERAGE PERSON IS IN THE MIDDLE, COMPARED WITH OTHER PEOPLE. FOR INSTANCE, THE AVERAGE HEIGHT OF OUR FRIENDS HERE IS SOMEWHERE BETWEEN 54 INCHES, THE SHORTEST, AND 66 INCHES, THE TALLEST.

THE AVERAGE OF TWO NUMBERS "SPLITS THE DIFFERENCE": IT'S EXACTLY HALFWAY BETWEEN THEM. HERE THE DIFFERENCE IS 12; THE AVERAGE IS 60.

$$66 - 54 = 12$$
$$\tfrac{1}{2}(12) = 6$$
$$54 + 6 = 60$$
$$66 - 6 = 60$$

GIVEN ANY TWO NUMBERS H AND h WITH $H \geq h$, HALF THE DIFFERENCE IS $(H-h)/2$; THE AVERAGE, WRITTEN $\bar{h}$, IS h PLUS THE SPLIT DIFFERENCE, $h + (H-h)/2$. THIS CAN BE SIMPLIFIED, BECAUSE:

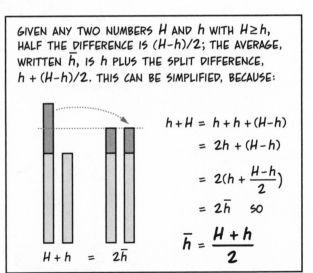

$$h + H = h + h + (H-h)$$
$$= 2h + (H-h)$$
$$= 2(h + \frac{H-h}{2})$$
$$= 2\bar{h} \quad \text{SO}$$

$$\bar{h} = \frac{H+h}{2}$$

$$H + h = 2\bar{h}$$

THE AVERAGE OF TWO NUMBERS IS **HALF THEIR SUM.** SIMILARLY, THE AVERAGE OF MANY NUMBERS $A_1, A_2, A_3, \ldots, A_n$, IS $1/n$ TIMES THEIR SUM. AGAIN WRITING THE AVERAGE AS $\bar{A}$,

$$\bar{A} = \frac{A_1 + A_2 + \ldots + A_n}{n}$$

THE AVERAGE HEIGHT OF ALL FIVE OF OUR HEROES, THEN, IS

$$\frac{66 + 66 + 64 + 60 + 54}{5}$$

$$= \frac{310}{5} = 62 \text{ INCHES}$$

NOTE THAT WE ADDED 66 TWICE, BECAUSE TWO DIFFERENT PEOPLE HAD THAT HEIGHT!

WE CAN ALSO ASK ABOUT SEPARATE HEIGHT AVERAGES FOR MALES AND FEMALES. THESE ARE:

MAYBE I CAN RAISE THE AVERAGE THIS WAY...

MALES
$$\frac{66 + 66 + 60}{3} = 64"$$

FEMALES
$$\frac{64 + 54}{2} = 59"$$

FINALLY, WHAT HAPPENS WHEN WE AVERAGE THE **MALE AND FEMALE AVERAGES?** LET'S SEE...

$$\frac{64 + 59}{2} = \frac{125}{2} = 62.5$$

CLOSE, BUT **NOT 62!!!** COMBINING ONE GROUP'S AVERAGE WITH ANOTHER GROUP'S AVERAGE GIVES A **DIFFERENT RESULT** FROM THE AVERAGE OF EVERYONE TAKEN TOGETHER!!

ANYONE BEGINNING TO SEE P*****'S MISTAKE?

NO!

159

Weights

NOW LET'S FORGET ABOUT HEIGHTS AND SIMPLY THINK OF TWO POINTS A AND B ON THE NUMBER LINE. THE AVERAGE $(A+B)/2$, HALFWAY BETWEEN THE POINTS, IS WHERE THE LINE SEGMENT BETWEEN A AND B WOULD **BALANCE** LIKE A SEE-SAW.

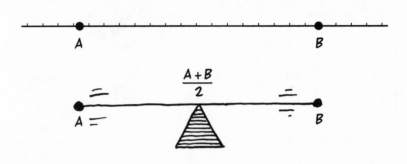

IT WOULD BALANCE, THAT IS, IF THERE WERE **EQUAL WEIGHTS** AT EACH END. BUT WHAT IF THE WEIGHTS ARE **DIFFERENT**?

OW!

IF THE WEIGHTS ARE DIFFERENT, THE BALANCE POINT HAS TO BE CLOSER TO THE HEAVIER END, AS YOU MAY KNOW FROM PLAYGROUND EXPERIENCE. WHERE IS THIS POINT?

LUCKILY, THE SEESAW IS DESCRIBED BY A SIMPLE EQUATION THAT WILL LOCATE THIS BALANCING POINT OR **CENTER OF GRAVITY.** IF W_A IS THE WEIGHT AT A, W_B THE WEIGHT AT B, L_A THE LENGTH OF A'S SIDE, AND L_B THE LENGTH OF B'S SIDE, THEN

$$W_A L_A = W_B L_B$$

IN BALANCE, THESE PRODUCTS ARE EQUAL. IF THE WEIGHT W_A GOES UP, ITS DISTANCE L_A MUST GO DOWN IN ORDER TO KEEP THE PRODUCT $W_A L_A$ THE SAME.

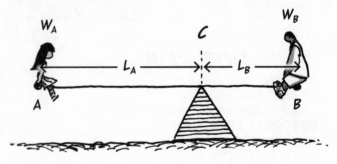

Example 1.

LET'S USE THE SEESAW EQUATION TO FIND A CENTER OF GRAVITY. SUPPOSE THE NUMBERS ARE $A = 3$ AND $B = 9$. IF $W_A = 75$ LBS SITS ON 3, AND $W_B = 150$ LBS SITS ON 9, WHERE IS THE CENTER OF GRAVITY C?

SOMEWHERE IN THERE...

SOLUTION: THE LENGTH $L_A = C - 3$; THE LENGTH $L_B = 9 - C$. PUTTING THAT INTO THE SEESAW EQUATION, WE CAN SOLVE FOR C.

$$W_A L_A = W_B L_B$$

$$75(C - 3) = 150(9 - C)$$

$$C - 3 = 2(9 - C) \quad \text{DIVIDING BY 75}$$

$$C - 3 = 18 - 2C$$

$$3C = 21$$

$$C = 7$$

FOLLOWING THE SAME STEPS FOR **ANY** NUMBERS $A \leq B$ AND WEIGHTS W_A AND W_B, WE CAN FIND THE CENTER OF GRAVITY C. FIRST NOTE THAT $L_A = C - A$ AND $L_B = B - C$; THEN...

$$W_A L_A = W_B L_B$$

$$W_A(C - A) = W_B(B - C)$$

$$W_A C + W_B C = W_A A + W_B B$$

$$C(W_A + W_B) = W_A A + W_B B$$

SO...

$$C = \frac{W_A A + W_B B}{W_A + W_B}$$

THIS POINT IS ALSO CALLED THE **WEIGHTED AVERAGE** OF A AND B, WITH THE WEIGHTS W_A AT A AND W_B AT B.

Example 1 Again.

NOW WE CAN SIMPLY PLUG THE NUMBERS FROM EXAMPLE 1 INTO THE FORMULA FOR C. OF COURSE WE GET THE SAME ANSWER!

$$C = \frac{(75)(3) + (150)(9)}{75 + 150}$$

$$= \frac{225 + 1350}{225}$$

$$= 7$$

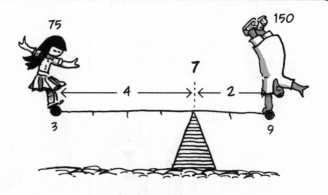

AND WE CHECK (AS WE DIDN'T THE FIRST TIME): $L_A = 4$, $L_B = 2$, AND THE SEESAW EQUATION IS SATISFIED.

$$(4)(75) = (2)(150) = 300$$

LET'S PLAY AROUND WITH THE WEIGHTED AVERAGE FORMULA SOME MORE TO GET A BETTER FEEL FOR IT—AND ALSO TO SIMPLIFY CALCULATIONS. FOR SIMPLICITY, WRITE W FOR THE SUM OF THE WEIGHTS:

$$W = W_A + W_B$$

NOW WORK ON THE FORMULA.

$$C = \frac{W_A A + W_B B}{W}$$

$$= \frac{W_A}{W} A + \frac{W_B}{W} B$$

THOSE FRACTIONS... THERE'S SOMETHING ABOUT THEM...

THOSE COEFFICIENTS W_A/W AND W_B/W ARE SPECIAL: **THEY ADD TO 1.**

$$\frac{W_A}{W} + \frac{W_B}{W} = \frac{W_A + W_B}{W}$$

$$= \frac{W}{W}$$

$$= 1 \ !$$

FOR INSTANCE?

IN EXAMPLE 1, THESE RATIOS WERE

$$\frac{W_A}{W} = \frac{75}{225} = \frac{1}{3}$$

$$\frac{W_B}{W} = \frac{150}{225} = \frac{2}{3}$$

AND THE WEIGHTED AVERAGE OF 3 AND 9 WITH WEIGHTS 75 AND 150 IS SUDDENLY VERY SIMPLE!

WHOA!

$$C = \frac{1}{3}(3) + \frac{2}{3}(9)$$

$$= 1 + 6 = 7$$

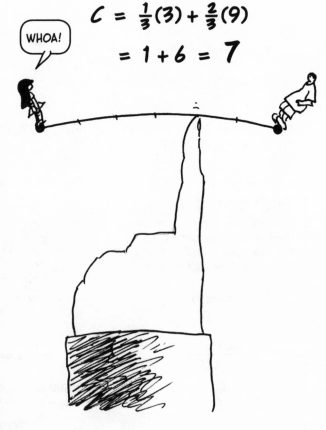

THIS MEANS: THE WEIGHTED AVERAGE DEPENDS NOT ON THE **VALUE** OF THE WEIGHTS, BUT ON THEIR **FRACTION OF THE TOTAL WEIGHT.** AS LONG AS THESE FRACTIONS ARE THE SAME, THE WEIGHTED AVERAGE IS THE SAME!

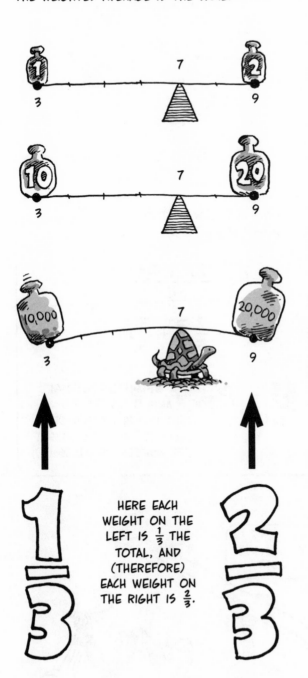

HERE EACH WEIGHT ON THE LEFT IS $\frac{1}{3}$ THE TOTAL, AND (THEREFORE) EACH WEIGHT ON THE RIGHT IS $\frac{2}{3}$.

NOW WE CAN THINK OF A WEIGHTED AVERAGE OF A AND B AS A SUM

WHERE $p + q = 1$. (THINK OF 1/3 AND 2/3, 1/4 AND 3/4, 2/5 AND 3/5, ETC.).

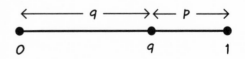

THIS NUMBER $pA + qB$ IS "q OF THE WAY FROM A TO B," AS IN, WHEN B IS WEIGHTED BY $\frac{2}{3}$, THEN THE WEIGHTED AVERAGE IS $\frac{2}{3}$ OF THE WAY FROM A TO B. ALGEBRAICALLY, START AT A AND ADD $q(B - A)$ TO GET $C = A + q(B - A)$.

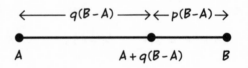

BECAUSE

$$A + q(B - A)$$

$$= (1 - q)A + qB$$

$$= pA + qB \quad \text{(SUBSTITUTING } p = 1 - q)$$

IN EXAMPLE 1, YOU CAN SEE THAT 7 IS EXACTLY $\frac{2}{3}$ OF THE WAY FROM 3 TO 9.

THE WEIGHTED AVERAGE IS ALWAYS CLOSER TO THE "HEAVIER" END.

OKAY... IS THERE ANY USE FOR WEIGHTED AVERAGES, ASIDE FROM BALANCING SEESAWS? YES! **WE MUST USE WEIGHTED AVERAGES WHEN AVERAGING RATES OR OTHER AVERAGES.**

I ALWAYS SAID MATH IS A PLAY-GROUND!

Example 2. GOING BACK TO AVERAGE HEIGHTS ON P. 159, LET'S WRITE $\bar{F}$ FOR THE FEMALE AVERAGE AND $\bar{M}$ FOR THE MALE AVERAGE. WE HAVE

$$\bar{F} = \frac{64 + 54}{2} \qquad \bar{M} = \frac{66 + 66 + 60}{3}$$

SO

$$64 + 54 = 2\bar{F} \qquad 66 + 66 + 60 = 3\bar{M}$$

THE OVERALL AVERAGE HEIGHT $\bar{H}$ IS

$$\bar{H} = \frac{64 + 54 + 66 + 66 + 60}{5}$$

BUT WE JUST SAW THAT $64 + 54 = 2\bar{F}$ AND $66 + 66 + 60 = 3\bar{M}$, SO

$$\bar{H} = \frac{2\bar{F} + 3\bar{M}}{5}$$

$$= \frac{2}{5}\bar{F} + \frac{3}{5}\bar{M}$$

$\bar{H}$ IS A **WEIGHTED** AVERAGE OF $\bar{F}$ AND $\bar{M}$, WHERE $\bar{F}$'S WEIGHT IS THE NUMBER OF FEMALES (2) AND $\bar{M}$'S WEIGHT IS THE NUMBER OF MALES (3).

WE CAN CHECK THIS.

$$\frac{2}{5}(59) + \frac{3}{5}(64)$$

$$= 23.6 + 38.4$$

$$= \mathbf{62}$$

EXACTLY THE OVERALL AVERAGE.

JUST A COINCIDENCE, RIGHT?

MORE **Examples:**

3. A CAR GOES FORWARD 60 MI/HR FOR 4 HOURS, THEN SPEEDS UP AND GOES 70 MI/HR FOR 2 HOURS. WHAT'S THE AVERAGE SPEED $\bar{v}$ OVER THE FULL 6 HOURS?

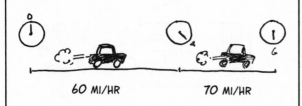

60 MI/HR   70 MI/HR

SOLUTION: $\bar{v}$ IS THE TOTAL DISTANCE d DIVIDED BY TIME TOTAL t.

$$d = (60 \text{ MI/HR})(4 \text{ HR}) + (70 \text{ MI/HR})(2 \text{ HR})$$

$$t = 4 \text{ HR} + 2 \text{ HR}$$

$$\bar{v} = \frac{(4)(60)+(2)(70)}{6} = \mathbf{63\tfrac{2}{3}} \text{ MI/HR}$$

THIS IS THE WEIGHTED AVERAGE OF THE SPEEDS: EACH SPEED IS WEIGHTED BY THE **AMOUNT OF TIME** SPENT AT THAT SPEED.

4. A BATTER HITS .330 IN HIS FIRST 100 AT-BATS AND .285 IN HIS NEXT 200 AT-BATS. WHAT WAS HIS OVERALL BATTING AVERAGE? **SOLUTION:** HIS BATTING AVERAGE (B.A.) IS TOTAL HITS DIVIDED BY TOTAL AT-BATS, OR

$$\text{B.A.} = \frac{(100)(.330) + (200)(.285)}{100 + 200} \quad \begin{matrix} \leftarrow \text{TOTAL HITS} \\ \leftarrow \text{TOTAL AT-BATS} \end{matrix}$$

IT'S ANOTHER WEIGHTED AVERAGE. EACH PARTIAL BATTING AVERAGE IS WEIGHTED BY ITS **NUMBER OF AT-BATS.** THIS WORKS OUT TO $\frac{1}{3}(.330) + \frac{2}{3}(.285) = \mathbf{.300}.$

IN GENERAL, IF SOMETHING HAPPENS AT RATE r_1 FOR TIME t_1, THEN CHANGES RATES TO r_2 FOR TIME t_2, THE **OVERALL AVERAGE RATE** $\bar{r}$ FOR THE FULL TIME IS THE **WEIGHTED AVERAGE** OF r_1 AND r_2. EACH RATE IS WEIGHTED BY THE TIME IT WAS IN EFFECT.

$$\bar{r} = \frac{r_1 t_1 + r_2 t_2}{t_1 + t_2}$$

BY THE WAY, THE ORDINARY AVERAGE $(A + B)/2$ IS A WEIGHTED AVERAGE—WITH EQUAL WEIGHTS! IT'S

$$\frac{1}{2}A + \frac{1}{2}B$$

WE CAN ALSO HAVE A WEIGHTED AVERAGE OF **MANY** NUMBERS. IF A_1, A_2, ..., A_n ARE THE NUMBERS, AND w_1, w_2, ..., w_n ARE THE WEIGHTS, THEN THE WEIGHTED AVERAGE $\bar{A}$ IS THIS:

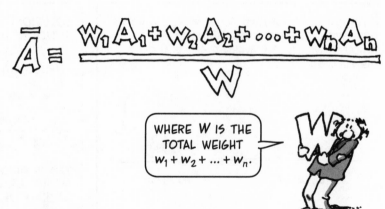

$$\bar{A} = \frac{w_1 A_1 + w_2 A_2 + \ldots + w_n A_n}{W}$$

WHERE W IS THE TOTAL WEIGHT $w_1 + w_2 + \ldots + w_n$.

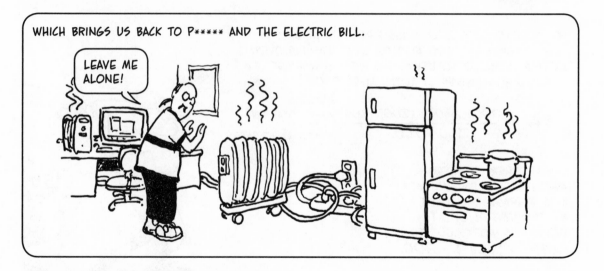

WHICH BRINGS US BACK TO P***** AND THE ELECTRIC BILL.

LEAVE ME ALONE!

P*****'S MISTAKE WAS TO IGNORE THE **AMOUNT OF EACH MONTHLY BILL.** HERE IS WHAT THE NUMBERS LOOKED LIKE, ROUNDED TO THE NEAREST WHOLE DOLLAR. P*****'S PERCENTAGE IS ON THE LEFT, THE AMOUNT OF THE TOTAL BILL IS IN THE MIDDLE, AND HIS SHARE IS ON THE RIGHT.

YEAH? SO?

	PCT		TOTAL BILL		P***** PAID
JUNE	.14	×	$117	=	$16
JULY	.17	×	$122	=	$21
AUGUST	.14	×	$96	=	$13
SEPT.	.25	×	$176	=	$44
OCT.	.26	×	$215	=	$56
NOV.	.30	×	$248	=	$74
DEC.	.28	×	$255	=	$71
TOTAL			**$1229**		**$295**

AT THIS POINT, THE EASIEST WAY TO FIND P*****'S **AVERAGE PERCENTAGE** IS TO DIVIDE THE TOTAL HE PAID IN 7 MONTHS BY THE TOTAL BILL OVER THE SAME TIME. THAT'S

$$\frac{\$295}{\$1229} = \mathbf{24\%}$$

NOT THE **22%** THAT P***** GOT WHEN HE AVERAGED THE NUMBERS IN COLUMN 1.

EHRRMM... WHY NOT?

BY NOW YOU SHOULD REALIZE THAT WHAT WE HAVE HERE IS A **WEIGHTED AVERAGE.** EACH MONTHLY PERCENTAGE IS WEIGHTED BY THAT MONTH'S **TOTAL BILL,** WHICH REFLECTS THE TOTAL AMOUNT OF ELECTRICITY USED BY THE WHOLE BUILDING. THERE WAS MORE USAGE IN THE WINTER MONTHS—THE SAME MONTHS WHEN P*****'S **PERCENTAGES WERE HIGHER.**

SEE? YOU JUST, UM, PLUG IN...

MORE USAGE **AND** HIGHER PERCENTAGES

$$\frac{(.14)(117)+(.17)(122)+(.14)(96)+(.25)(176)+(.26)(215)+(.3)(248)+(.28)(255)}{117+122+96+176+215+248+255}$$

WHY? BILLS USUALLY GO UP IN WINTER, WHEN IT'S DARKER AND COLDER. IN ADDITION, THERE WAS A **DIFFERENCE** BETWEEN P***** AND THE OTHER TENANTS: P***** **LIVED IN THE BUILDING,** WHILE THE REST OF US ONLY WORKED THERE... SO, AT NIGHT, WHEN THE REST OF US WERE AWAY, P***** RAN HIS LIGHTS AND ELECTRIC HEATER, AND UP WENT HIS PERCENTAGE... AND SO ENDS THE STORY OF PIG-HEADED P*****.

22%, NOT A PENNY MORE!

Problems

1. FIND THE (UNWEIGHTED) AVERAGE OF THE NUMBERS:

a. 7 AND 17

b. 9 AND 12

c. 1,000,000 AND 1,000,002

d. −9 AND −12

e. 9 AND −12

f. 55 AND −55

g. −1,000,000 AND 1,000,002

h. 19, 21, 23

i. 5, 38, 2

j. 103, 4, −100, 1

2. FIND THE WEIGHTED AVERAGES OF

a. 7 WITH WEIGHT 3 AND 11 WITH WEIGHT 1

b. 1 WITH WEIGHT 2 AND 2 WITH WEIGHT 1

c. −2 WITH WEIGHT 5 AND 2 WITH WEIGHT 15

d. 0 WITH WEIGHT 11 AND 12 WITH WEIGHT 1

e. 0 WITH WEIGHT 0 AND A WITH WEIGHT w

f. 0 WITH WEIGHT 3 AND −1 WITH WEIGHT 9

g. 100 WITH WEIGHT .23 AND 1,000 WITH WEIGHT .77

3. GIVEN ANY FOUR NUMBERS a, b, c, d, SHOW THAT

$$\text{IF } \frac{a}{a+b} = \frac{c}{c+d} \quad \text{THEN} \quad \frac{a}{b} = \frac{c}{d}$$

4. DRAW THE POINT ON THE LINE BETWEEN A AND B INDICATED BY THE EXPRESSION.

A ●————————————● B

a. $\frac{1}{10}A + \frac{9}{10}B$

b. $\frac{1}{4}A + \frac{3}{4}B$

c. $\frac{2}{3}A + \frac{1}{3}B$

d. $\frac{999}{1000}A + \frac{1}{1000}B$

e. $\frac{3A + 2B}{5}$

f. $\frac{610A + 305B}{915}$

5. KEVIN IS HANGING WEIGHTS FROM A STICK, WHICH ITSELF HANGS FROM A STRING. HE ATTACHES A 7-OUNCE PIECE 3 INCHES FROM THE STRING, AND A 1-OUNCE PIECE 9 INCHES FROM THE STRING ON THE OTHER SIDE. THE THIRD PIECE WEIGHS 3 OUNCES. WHERE MUST HE HANG IT TO MAKE THE MOBILE BALANCE? IGNORE THE WEIGHT OF THE STICK AND STRING. (HINT: TAKE C, THE BALANCE POINT, TO BE 0.)

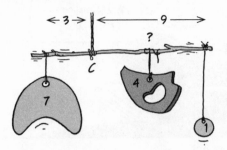

6. SUPPOSE A CAR GOES 40 MI/HR FOR 3 HOURS AND 80 MI/HR FOR 2 HOURS. WHAT'S THE AVERAGE SPEED FOR THE WHOLE 5-HOUR TRIP?

7. CELIA TAKES A TRIP TO VISIT HER COUSIN 120 MILES AWAY. GOING OUT, HER SPEED IS 40 MI/HR; COMING BACK, IT'S 60 MI/HR. WHAT'S HER AVERAGE SPEED FOR THE ROUND TRIP? (HINT: HOW MUCH TIME WAS SPENT IN EACH DIRECTION?)

8. IF A CAR GOES AT A SPEED v_1 FOR A DISTANCE d_1 AND THEN GOES v_2 FOR A DISTANCE d_2, FIND AN EXPRESSION IN d_1, d_2, v_1, AND v_2 FOR THE AVERAGE SPEED OVER THE WHOLE TRIP.

9. MOMO GETS 4 AT-BATS IN THE FIRST HALF OF THE SEASON AND HITS .750. IN THE SECOND HALF OF THE SEASON, SHE GETS 92 AT-BATS AND HITS .290. WHAT'S HER BATTING AVERAGE FOR THE SEASON?

10. THAT SAME SEASON, JESSE'S FIRST-HALF BATTING AVERAGE WAS LESS THAN MOMO'S, AND HIS SECOND-HALF BATTING AVERAGE WAS ALSO LESS THAN MOMO'S SECOND-HALF AVERAGE. IS IT POSSIBLE FOR HIM TO HAVE HAD A HIGHER BATTING AVERAGE FOR THE FULL SEASON?

Chapter 13
Squares

TO **SQUARE** A NUMBER MEANS TO MULTIPLY IT BY ITSELF, LIKE THIS:

$$5^2 = 25.$$

WE CAN ALSO SQUARE A VARIABLE, LIKE THIS:

$$x^2$$

JUST TO REMIND YOU, IT'S CALLED "SQUARING" x BECAUSE x^2 IS THE AREA OF A SQUARE WITH ALL ITS SIDES EQUAL TO x.

EXPRESSIONS CONTAINING SQUARES OF VARIABLES (OR PRODUCTS OF TWO DIFFERENT VARIABLES) SUCH AS $4x^2 - 3xy + y^2$, ARE CALLED **QUADRATIC**, FROM THE LATIN *QUADRA*, MEANING "SQUARE."

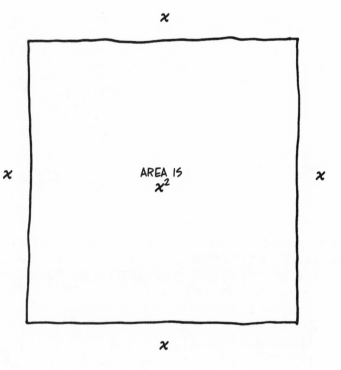

x

x x

AREA IS
x^2

x

WHAT ELSE WOULD YOU CALL IT?

169

THE OLDEST KNOWN QUADRATIC QUESTION IS A 4,000-YEAR-OLD BABYLONIAN PUZZLER: GIVEN THE TOTAL DISTANCE AROUND A RECTANGULAR FIELD, AND THE FIELD'S AREA, HOW LONG ARE ITS SIDES? FOR INSTANCE, IF THE PERIMETER (THE DISTANCE AROUND) IS 32, AND THE AREA IS 63, FIND TWO NUMBERS r AND s SO THAT $2r+2s=32$ AND $rs=63$.

$$2r+2s = 32$$
$$rs = 63$$

THAT PRODUCT rs IS A CLUE THAT WE'RE IN QUADRATIC TERRITORY HERE...

ALGEBRA? THE FIELD IS WIDE OPEN...

ANOTHER OLDIE—AND ONE OF THE COOLEST OF ALL QUADRATIC RELATIONSHIPS—COMES FROM THE ANCIENT GREEK **PYTHAGORAS.** PYTHAGORAS SHOWED HOW TO EXPRESS THE **DISTANCE** BETWEEN TWO POINTS ON A PLANE IN TERMS OF THE RISE AND THE RUN BETWEEN THEM. IF x IS THE RUN, y IS THE RISE, AND r IS THE DISTANCE, THEN THEY SATISFY A SIMPLE FORMULA:

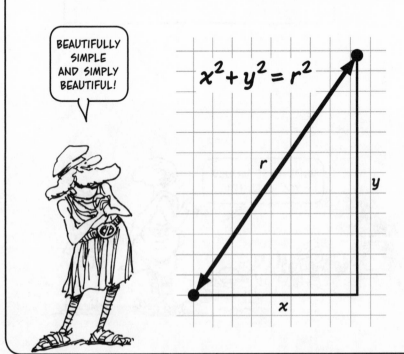

BEAUTIFULLY SIMPLE AND SIMPLY BEAUTIFUL!

$$x^2 + y^2 = r^2$$

(YOU'LL SEE **WHY** IT'S TRUE WHEN YOU TAKE GEOMETRY, BUT IT IS **NEVER** TOO EARLY TO LEARN THIS FABULOUS FORMULA!!!)

AND THEN THERE'S **BALLISTICS**, OR AIMING CANNONBALLS. IT TURNS OUT THAT YOU CAN EXPRESS THE BALL'S HEIGHT h IN TERMS OF ITS (HORIZONTAL) DISTANCE s FROM THE CANNON LIKE SO:

$$h = as^2 + bs + h_0$$

HERE h_0 IS THE HEIGHT OF THE CANNON ITSELF, WHILE a AND b ARE NUMBERS DEPENDING ON THE TILT OF THE GUN AND THE SPEED OF THE BALL AS IT EXITS THE MUZZLE.

WHEN THE CANNONBALL HITS THE GROUND, $h = 0$, AND THE PROBLEM IS TO FIND THE VALUE OF s AT THAT POINT. IN OTHER WORDS, THE GUNNER HAS TO SOLVE THIS EQUATION FOR s:

$$as^2 + bs + h_0 = 0$$

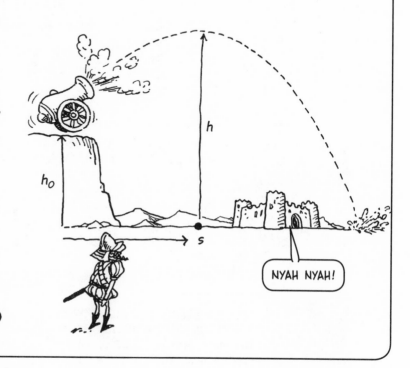

NYAH NYAH!

JUST IMAGINE THE INTEREST! AND SO, NOT TOO LONG AFTER CANNON ARRIVED IN EUROPE, QUADRATIC EQUATIONS FOLLOWED...

171

OUR FIRST QUADRATIC WILL BE...

The Expression (x+r)(x+s)

BY NOW, WE HAVE SEEN MANY
EXPRESSIONS LIKE $a(c+d)$ AND
$b(c+d)$. WHAT IS THEIR SUM?

$$a(c+d) + b(c+d) = ?$$

REGARDING $c+d$ AS A SINGLE
NUMBER, WE CAN USE THE
DISTRIBUTIVE LAW TO PULL
THAT FACTOR OUT OF THE SUM:

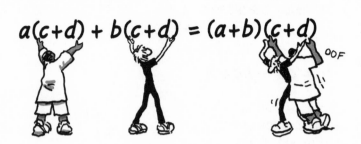

$$a(c+d) + b(c+d) = (a+b)(c+d)$$

SO $a(c+d) + b(c+d) = (a+b)(c+d)$. WE ALSO KNOW THAT $a(c+d) + b(c+d) = ac + ad + bc + bd$.
PUTTING THESE TWO TOGETHER GIVES US THE **EXPANSION OF** $(a+b)(c+d)$:

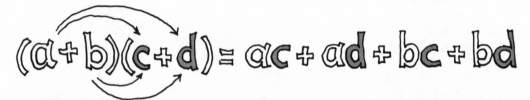

$$(a+b)(c+d) = ac + ad + bc + bd$$

MULTIPLY EVERY POSSIBLE PRODUCT
CONSISTING OF ONE FACTOR FROM
THE FIRST SUM AND ONE FROM THE
SECOND SUM, THEN ADD 'EM UP!

YOU CAN DRAW $(a+b)(c+d)$ AS A
RECTANGLE, ITS TWO SIDES BEING
$a+b$ AND $c+d$. THE TOTAL AREA,
$(a+b)(c+d)$, IS THE SUM OF THE
FOUR SMALLER BOXES' AREAS.

	a	b	
	ac	bc	c
	ad	bd	d

NOW SUPPOSE r AND s ARE ANY NUMBERS. USING WHAT WE JUST LEARNED, WE CAN EXPAND $(x+r)(x+s)$.

$$(x+r)(x+s)$$

$$= xx + rx + sx + rs$$

$$= x^2 + (r+s)x + rs$$

THE RESULTING QUADRATIC EXPRESSION IN x HAS A CONSTANT TERM EQUAL TO THE PRODUCT rs, AND A "LINEAR COEFFICIENT," THE COEFFICIENT OF x, EQUAL TO THE SUM $r+s$.

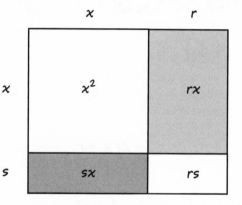

THE SHADED AREA IS $rx + sx = (r+s)x$.

Examples:

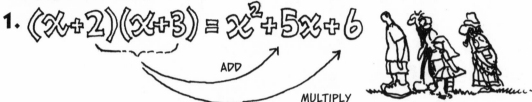

1. $(x+2)(x+3) = x^2 + 5x + 6$

ADD

MULTIPLY

2. $(x+1)(x+7)$
$= x^2 + (1+7)x + (1)(7)$
$= x^2 + 8x + 7$

EXAMPLES 3–5 SHOW THAT r AND s NEED NOT BE POSITIVE.

3. $(x-1)(x+2)$
$= x^2 + (2-1)x + (-1)(2)$
$= x^2 + x - 2$

4. $x(x+3) = x^2 + 3x$

(HERE $r = 0$.)

5. $(x-1)(x-3)$
$= x^2 + (-1-3)x + (-1)(-3)$
$= x^2 - 4x + 3$

BY THE WAY, DID YOU RECOGNIZE THE "BABYLONIAN NUMBERS" THAT POPPED UP HERE IN THE COEFFICIENTS rs AND $r+s$? (WELL, ACTUALLY, $r+s$ IS ONLY HALF THE BABYLONIAN SUM $2r+2s$, BUT THAT'S NO BIG DEAL.) HOW ABOUT THAT?

CIVILIZATIONS GET OLD, BUT MATH—NEVER!

Two SPECIAL CASES

$(x+r)^2$

WHEN WE **SQUARE** THE LINEAR EXPRESSION $(x+r)$, THE RESULT HAS A BEAUTIFUL PATTERN:

$$(x+r)^2 = x^2 + 2rx + r^2$$

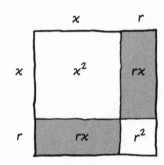

THE TWO SHADED AREAS ADD UP TO... HMMM... $rx + rx$...

$2rx$?

Example 6. THESE REALLY ARE ADORABLE, AREN'T THEY?

$(x+1)^2 = x^2 + 2x + 1$
$(x+2)^2 = x^2 + 4x + 4$
$(x+3)^2 = x^2 + 6x + 9$
$(x+4)^2 = x^2 + 8x + 16$

SQUARES WITH NEGATIVE r ARE PRETTY CUTE TOO...

$(x-1)^2 = x^2 - 2x + 1$
$(x-2)^2 = x^2 - 4x + 4$
$(x-3)^2 = x^2 - 6x + 9$
$(x-4)^2 = x^2 - 8x + 16$

$(x+r)(x-r)$

THIS ONE MAGICALLY GETS RID OF THE MIDDLE TERM, BECAUSE $r + (-r) = 0$. THE CONSTANT TERM IS $(r)(-r) = -r^2$.

$$(x+r)(x-r) = x^2 - r^2$$

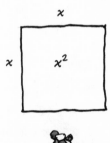

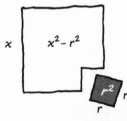

Example 7. WHEN $r=1$, THIS BECOMES ANOTHER BEAUTIFUL FORMULA:

$$x^2 - 1 = (x+1)(x-1)$$

AND ALSO

$x^2 - 4 = (x+2)(x-2)$
$x^2 - 9 = (x+3)(x-3)$

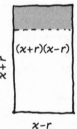

MENTAL ARITHMETIC Trick:

THE EQUATION $(x+1)(x-1) = x^2 - 1$ OPENS A SHORTCUT FOR MULTIPLYING NUMBERS THAT DIFFER BY 2.

Example 8. TO MULTIPLY 15 × 17, WE FIRST NOTICE THAT 15 = 16 − 1 AND 17 = 16 + 1, SO

$$15 \times 17 = (16-1)(16+1) = 16^2 - 1 = 256 - 1$$

$$= \mathbf{255}$$

TO DO THESE PRODUCTS IN YOUR HEAD, YOU'LL NEED TO MEMORIZE SOME SQUARES; THIS TABLE WILL GIVE YOU A START.

n	n^2
1	1
2	4
3	9
4	16
5	25
6	36
7	49
8	64
9	81
10	100
11	121
12	144
13	169
14	196
15	225
16	256
17	289
18	324
19	361
20	400
21	441
22	484
23	529
24	576
25	625
26	676
27	729
28	784
29	841
30	900
31	961
32	1,024
33	1,089

THE TRICK WORKS FOR ANY PAIR OF NUMBERS THAT DIFFER BY A SMALL EVEN NUMBER. SPLIT THE DIFFERENCE AND USE THE FORMULA.

I LOVE ALL TABLES!

Example 9. FIND 98 × 102.

THE NUMBER 100 IS HALFWAY BETWEEN THE TWO FACTORS.

$$98 = 100 - 2, \quad 102 = 100 + 2, \text{ SO}$$

$$98 \times 102 = 100^2 - 2^2$$

$$= 10,000 - 4$$

$$= \mathbf{9,996}$$

ROOTS of an Expression

THE **ROOTS** OF AN EXPRESSION ARE THE NUMBERS WHERE ITS VALUE IS **ZERO**. IN SYMBOLS, r IS A ROOT OF THE EXPRESSION $ax^2 + bx + c$ IF $ar^2 + br + c = 0$.

THAT IS, A ROOT OF $ax^2 + bx + c$ IS ANY SOLUTION OF THE EQUATION

$$ax^2 + bx + c = 0$$

ROOTS ARE VALUES OF THE VARIABLE THAT "ZERO OUT" THE EXPRESSION. AS WE'LL SEE, A QUADRATIC EXPRESSION GROWS OUT OF ITS ROOTS SOMEHOW...

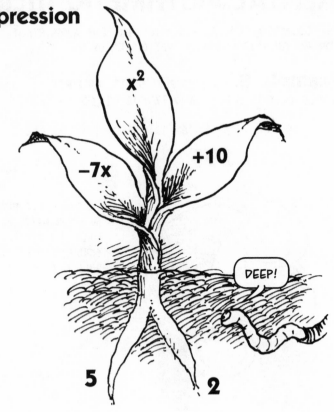

... AND OUR GOAL IS TO DIG UP ROOTS.

Example 10.

-2 IS A ROOT OF THE EXPRESSION $3x^2 + 15x + 18$, BECAUSE WHEN WE PLUG IN -2 FOR x AND EVALUATE THE EXPRESSION, WE GET ZERO.

$$3(-2)^2 + (15)(-2) + 18$$
$$= (3)(4) - 30 + 18$$
$$= 12 - 30 + 18 = 0$$

YEAH, BUT WHERE DO YOU DIG UP -2 IN THE FIRST PLACE?

THAT IS THE QUESTION!

Important NOTE:

GIVEN AN EQUATION LIKE $3x^2 + 15x + 18 = 0$, WE CAN DIVIDE BOTH SIDES BY ITS "LEADING COEFFICIENT," THE COEFFICIENT OF x^2, IN THIS CASE 3, AND THE EQUATION IS STILL TRUE.

$$3x^2 + 15x + 18 = 0$$
$$x^2 + 5x + 6 = 0$$

YOU CAN CHECK THAT -2 IS A ROOT OF $x^2 + 5x + 6$, AND ALSO THAT -3 IS A ROOT OF BOTH!

EITHER EQUATION IS TRUE IF THE OTHER ONE IS; THAT IS, THEY HAVE THE SAME SOLUTIONS... OR, IN OUR NEW LANGUAGE OF ROOTS, **THE EXPRESSION $3x^2 + 15x + 18$ HAS THE SAME ROOTS AS $x^2 + 5x + 6$.**

WE CAN DO THIS WITH ANY QUADRATIC EQUATION. THE EQUATION $ax^2 + bx + c = 0$ HAS THE SAME SOLUTIONS AS

$$x^2 + \frac{b}{a} + \frac{c}{a}$$

AS FAR AS FINDING ROOTS IS CONCERNED, THEN, WE CAN ASSUME THAT AN EXPRESSION'S **LEADING COEFFICIENT IS 1.**

The Roots of (x−r)(x−s)

ON PAGE 173, WE SAW HOW TO EXPAND $(x+r)(x+s)$. IF WE CHANGE THOSE PLUS SIGNS TO MINUSES, WE FIND THAT $(x-r)(x-s)$ EXPANDS IN MUCH THE SAME WAY.

$$(x-r)(x-s) = x^2 - rx - sx + (-r)(-s)$$
$$= x^2 - (r+s)x + rs$$

WE SAW ONE LIKE THIS IN EXAMPLE 5. HERE'S ANOTHER:

Example 11.

$$(x-4)(x-7) = x^2 - (4+7)x + (4)(7)$$
$$= x^2 - 11x + 28$$

THE ROOTS OF $(x-r)(x-s)$ ARE SITTING RIGHT IN FRONT OF US: THEY'RE r AND s!! SUBSTITUTING $x = r$ MAKES THE FIRST FACTOR $r - r = 0$, SO THE PRODUCT IS ZERO; SIMILARLY, $x = s$ MAKES THE SECOND FACTOR ZERO.

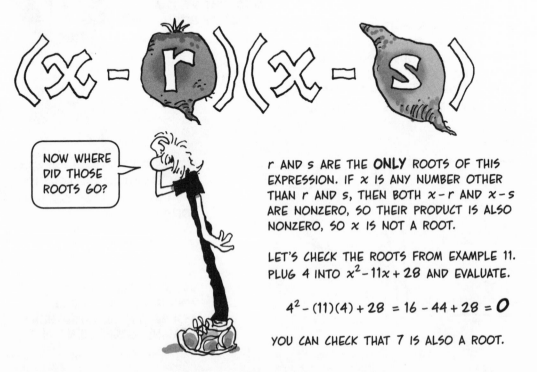

r AND s ARE THE **ONLY** ROOTS OF THIS EXPRESSION. IF x IS ANY NUMBER OTHER THAN r AND s, THEN BOTH $x - r$ AND $x - s$ ARE NONZERO, SO THEIR PRODUCT IS ALSO NONZERO, SO x IS NOT A ROOT.

LET'S CHECK THE ROOTS FROM EXAMPLE 11. PLUG 4 INTO $x^2 - 11x + 28$ AND EVALUATE.

$$4^2 - (11)(4) + 28 = 16 - 44 + 28 = 0$$

YOU CAN CHECK THAT 7 IS ALSO A ROOT.

THIS IS WHAT I MEANT EARLIER WHEN I SAID THAT QUADRATIC EXPRESSIONS GROW FROM THEIR ROOTS. WE ARE OFTEN GIVEN AN EXPRESSION $x^2 + bx + c$ WITH ITS COEFFICIENTS 1, b, AND c, WHILE THE ROOTS r AND s REMAIN HIDDEN. IF WE CAN FIND THEM, THEN WE'LL KNOW THAT OUR EXPRESSION WAS "REALLY" THE PRODUCT $(x-r)(x-s)$.

CAN YOU DIG IT?

$$x^2 + bx + c$$
$$r + s = -b$$
$$rs = c$$
$$(x-r)(x-s)$$

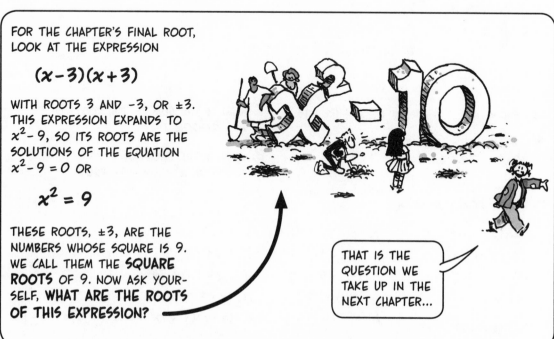

FOR THE CHAPTER'S FINAL ROOT, LOOK AT THE EXPRESSION

$$(x-3)(x+3)$$

WITH ROOTS 3 AND −3, OR ±3. THIS EXPRESSION EXPANDS TO $x^2 - 9$, SO ITS ROOTS ARE THE SOLUTIONS OF THE EQUATION $x^2 - 9 = 0$ OR

$$x^2 = 9$$

THESE ROOTS, ±3, ARE THE NUMBERS WHOSE SQUARE IS 9. WE CALL THEM THE **SQUARE ROOTS** OF 9. NOW ASK YOURSELF, **WHAT ARE THE ROOTS OF THIS EXPRESSION?**

THAT IS THE QUESTION WE TAKE UP IN THE NEXT CHAPTER...

Problems

1. IN THE RECTANGLE FROM P. 172 THAT ILLUSTRATES $(a+b)(c+d)$, COLOR IN THE PARTS OF IT THAT ADD UP TO $a(c+d)$; $b(c+d)$; $(a+b)c$.

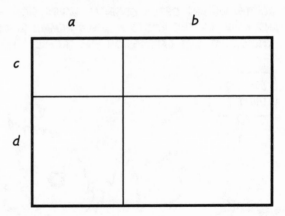

2. EXPAND BY MULTIPLYING:

   a. $(a+2)(b+3)$

   b. $x(x+5)$

   c. $3x(2x-3)$

   d. $(t-4)(t+4)$

   e. $(x-7)^2$

   f. $(7p-4)(2p-3)$

   g. $(3-x)(2-x)$

   h. $(x-5)(x+3)$

   i. $(t+3)^2$

   j. $(2x+3)(4x-5)$

   k. $7(p-1)(2p+5)$

3. QUICKLY CALULATE **a.** 12×14 **b.** 13×17

4. EXPRESS EACH PRODUCT AS A DIFFERENCE OF SQUARES, AND EVALUATE.

   a. $999 \times 1,001$ **d.** 25×35

   b. $995 \times 1,005$ **e.** 0.95×1.05

   c. 18×22 **f.** $9,999,000 \times 10,001,000$

5. WRITE THE ROOTS OF EACH EXPRESSION.

   a. $(x-2)(x-5)$ **e.** $(x-1)^2$

   b. $(x-2)(x+5)$ **f.** $(x+6)^2$

   c. $(x+3)(x+1)$ **g.** $(x-1)(x+3)(x-5)$

   d. $(x+r)(x+s)$

6a. SHOW THAT 3 IS A ROOT OF $x^2-8x+15$.

6b. SHOW THAT -7 IS A ROOT OF $2x^2+17x+21$.

7. WHAT IS THE **SUM** OF THE ROOTS OF $x^2-2000x+1$?

8. WHAT IS THE **PRODUCT** OF THE ROOTS OF $x^2+3x-17,458$?

9. EXPAND BY MULTIPLYING:

   a. $(p^2+q)(4+q)$

   b. $(a^2-b)(a^2+b)$

   c. $(t+1)(t^2-t+1)$

   d. $(x+1)(\frac{x}{2}+\frac{2}{3})$

   e. $(x-\frac{1}{2})^2$

   f. $(t+3)^3$

   g. $(2x+1)^2$

   h. $(3x-5)^2$

   i. $(ax+r)^2$

   j. $(x+1)^3$

   k. $(x-1)^3$

   l. $(x-1)(x^2+x+1)$

   m. $(x-1)(x^3+x^2+x+1)$

   n. $(x-1)(x^4-x^3+x^2-x+1)$

   o. $(x-r)(x^5+rx^4+r^2x^3+r^3x^2+r^4x+r^5)$

Chapter 14
Square Roots

At the end of the last chapter, we wondered about the roots of $x^2 - 10$. These would be solutions of $x^2 - 10 = 0$, or

$$x^2 = 10$$

What number's square is 10? Nobody knows exactly! But that doesn't stop us from giving it a name—the **square root of 10**—and writing it down this way. ➡

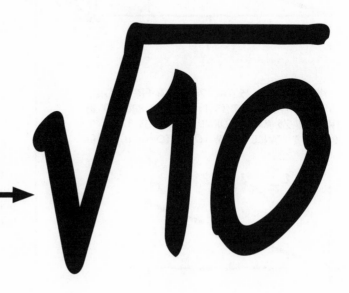

The symbol $\sqrt{}$ is called a **radical sign**. The word "radical," like "radish," comes from a latin root meaning... er... root.

ECCE RADIX!*

*BEHOLD THE ROOT.

FIRST LET ME CONVINCE YOU THAT THERE **IS** SUCH A NUMBER—BY DRAWING IT.

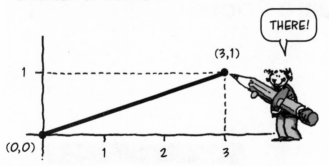

THERE!

THE NUMBER $\sqrt{10}$ IS THE DISTANCE FROM THE ORIGIN TO THE POINT (3,1). THIS IS PROVED BY PYTHAGORAS'S MAGIC FORMULA (SEE P. 170). IF r IS THE DISTANCE FROM (0,0) TO (x,y), THEN

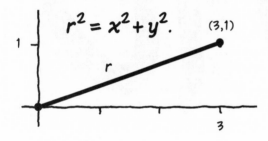

$$r^2 = x^2 + y^2.$$

HERE $r^2 = 3^2 + 1^2 = 9 + 1 = 10$, SO

$$r = \sqrt{10}$$

IF YOU PIVOT THE LINE SEGMENT AROUND THE ORIGIN DOWN TO THE x-AXIS, YOU CAN SEE THAT $\sqrt{10}$ FALLS A LITTLE BEYOND 3.

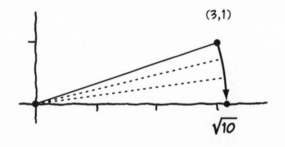

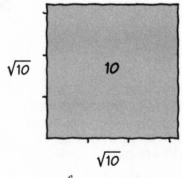

NOTE: $\sqrt{10}$ IS ALSO THE SIDE OF A SQUARE WITH AREA 10.

THIS NUMBER, $\sqrt{10}$, IS SLIGHTLY GREATER THAN 3.1622 AND SLIGHTLY LESS THAN 3.1623.

$$3.1622^2 = 9.99950884$$
$$3.1623^2 = 10.00014129$$

MY COMPUTER CALCULATES $\sqrt{10}$ TO FOURTEEN DECIMAL PLACES AS

3.162 277 660 168 38

BUT EVEN THAT IS A HAIR TOO LARGE. WE CAN NEVER WRITE A COMPLETE DECIMAL EXPANSION, BECAUSE $\sqrt{10}$ IS AN **IRRATIONAL** NUMBER.

ALWAYS REMEMBER!

$$(\sqrt{n})^2 = n$$

OKAY...

HERE ARE A FEW SQUARE ROOTS. NO NEED TO MEMORIZE ALL THESE!!!

n	$\sqrt{n}$
1	1
2	1.41421356...
3	1.73205080...
4	2
5	2.23606797...
6	2.44948974...
7	2.645751311...
8	2.82842712...
9	3
10	3.16227766...
11	3.31662479...
12	3.46410161...
13	3.60555127...
14	3.74165738...
15	3.87298334...
16	4

That **OTHER** Square Root

A POSITIVE NUMBER'S SQUARE IS OBVIOUSLY POSITIVE: $3 \times 3 = 9$. A NEGATIVE NUMBER'S SQUARE IS ALSO POSITIVE: $(-3)(-3) = 9$. AND $0^2 = 0$. IN OTHER WORDS, **ALL SQUARES ARE NON-NEGATIVE. NO NEGATIVE NUMBER HAS A REAL SQUARE ROOT.**

JUST UNREAL!

ON THE OTHER HAND, EVERY **POSITIVE** NUMBER HAS **TWO** SQUARE ROOTS, ONE POSITIVE AND ONE NEGATIVE. "THE" SQUARE ROOT OF 9 IS ACTUALLY TWO NUMBERS, 3 AND −3. **THE SYMBOL** $\sqrt{n}$ **ALWAYS REFERS TO THE POSITIVE SQUARE ROOT** (OR ZERO, IF $n = 0$). THE NEGATIVE SQUARE ROOT IS WRITTEN $-\sqrt{n}$.

$$\sqrt{9} = 3 \qquad -\sqrt{9} = -3$$

BOTH ARE SQUARE ROOTS OF 9.

REMEMBER THIS, TOO!

$$(-\sqrt{n})^2 = n$$

ADDING Square Roots

WHEN ADDING TWO SQUARE ROOTS, THERE IS OFTEN NO WAY TO SIMPLIFY THE SUM, AT LEAST WHEN WE SEE DIFFERENT NUMBERS UNDER THE RADICAL SIGN. HERE ARE SOME EXPRESSIONS THAT MUST BE LEFT AS IS:

$$1 + \sqrt{2} \qquad \sqrt{3} - \sqrt{11}$$
$$\sqrt{x} + \sqrt{y}$$

ON THE OTHER HAND, WHEN THE **SAME** NUMBER APPEARS UNDER BOTH RADICAL SIGNS, WE'RE IN LUCK!

$$\sqrt{3} + \sqrt{3} = 2\sqrt{3}$$
$$\sqrt{n} + \sqrt{n} = 2\sqrt{n}$$
$$a\sqrt{n} + b\sqrt{n} = (a+b)\sqrt{n}$$

THIS IS NOTHING BUT THE DISTRIBUTIVE LAW. SQUARE ROOTS BEHAVE LIKE ANY OTHER QUANTITY.

EVEN RADICALS OBEY THE **LAW!**

VERY OBEDIENT!

Example 1. SIMPLIFY $3\sqrt{15} + 2\sqrt{3} + \sqrt{15} + 4\sqrt{3}$.

WE GROUP LIKE TERMS, SO THE EXPRESSION BECOMES

$$3\sqrt{15} + \sqrt{15} + 2\sqrt{3} + 4\sqrt{3}$$
$$= (3 + 1)\sqrt{15} + (2 + 4)\sqrt{3}$$
$$= 4\sqrt{15} + 6\sqrt{3}$$

MULTIPLYING Square Roots

MULTIPLYING SQUARE ROOTS IS EASY—AS LONG AS EVERYTHING IS **POSITIVE**.

I LIKE TO STAY POSITIVE!

THE RULE IS SIMPLE. IF a AND b ARE ANY NON-NEGATIVE NUMBERS, THEN THE PRODUCT OF ROOTS IS THE ROOT OF THE PRODUCT.

$$\sqrt{ab} = \sqrt{a}\sqrt{b}$$

THIS FOLLOWS FROM POWER LAW #3 ON P. 119: $(xy)^2 = x^2y^2$. IF WE SQUARE THE PRODUCT $\sqrt{a} \cdot \sqrt{b}$, THIS HAPPENS:

$$\left(\sqrt{a}\sqrt{b}\right)^2 = \left(\sqrt{a}\right)^2\left(\sqrt{b}\right)^2 = ab$$

SINCE THAT THING INSIDE THE PARENTHESES ON THE LEFT SQUARES TO ab (AND IT'S NON-NEGATIVE), IT MUST BE $\sqrt{ab}$.

$$\sqrt{a}\sqrt{b} = \sqrt{ab}$$

I CAN BE NON-NEGATIVE, TOO, SOMETIMES...

IF BOTH a AND b ARE NEGATIVE (SO THAT $ab > 0$), THEN NEITHER $\sqrt{a}$ NOR $\sqrt{b}$ IS REAL, AND THE RULE DOESN'T HOLD. IN THAT CASE,

$$\sqrt{ab} = \sqrt{-a}\sqrt{-b}$$

THIS AMOUNTS TO SAYING: IF a AND b ARE BOTH POSITIVE OR BOTH NEGATIVE, THEN

$$\sqrt{ab} = \sqrt{|a|}\sqrt{|b|}$$

Example 2. $\sqrt{15} = \sqrt{5}\sqrt{3}$

Example 3. $\sqrt{12} = \sqrt{4}\sqrt{3} = 2\sqrt{3}$

AND CHECK THIS OUT!

SQUARE FACTORS come out!

ACCORDING TO THE PRODUCT RULE, $\sqrt{a^2} = \sqrt{|a|}\,\sqrt{|a|} = (\sqrt{|a|})^2 = |a|$. THIS FORMULA IS SO HANDY, I'LL WRITE IT LARGE.

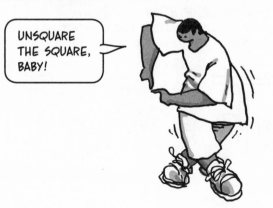

UNSQUARE THE SQUARE, BABY!

$$\sqrt{a^2} = |a|$$

WHEN $a \geq 0$, THIS IS SIMPLY

$$\sqrt{a^2} = a$$

WHICH LETS US PULL OUT ANY SQUARE FACTOR UNDER THE RADICAL SIGN (MAKING SURE TO UNSQUARE IT WHEN WE DO!):

THE REASON, AGAIN, IS THE PRODUCT RULE.

$$\sqrt{a^2 b} = \sqrt{a^2}\,\sqrt{b}$$
$$= |a|\sqrt{b}$$

$$\sqrt{a^2 b} = |a|\sqrt{b}$$

THIS LETS US SIMPLIFY THE SQUARE ROOT OF ANY NUMBER CONTAINING A SQUARE FACTOR.

Example 4. $\sqrt{63} = \sqrt{(9)(7)} = \sqrt{(3)^2(7)} = 3\sqrt{7}$

Example 5. $\sqrt{300} = \sqrt{(10)^2(3)} = 10\sqrt{3}$

Example 6. $\sqrt{3} + \sqrt{12} = \sqrt{3} + 2\sqrt{3} = 3\sqrt{3}$

Example 7. $\sqrt{2} + \sqrt{50} = \sqrt{2} + \sqrt{25 \cdot 2}$
$$= \sqrt{2} + 5\sqrt{2} = 6\sqrt{2}$$

IT PAYS TO ROOT, ROOT, ROOT...

QUOTIENTS of Roots

QUOTIENTS BEHAVE JUST LIKE PRODUCTS: THE QUOTIENT OF SQUARE ROOTS IS THE SQUARE ROOT OF THE QUOTIENT.

$$\sqrt{\dfrac{a}{b}} = \dfrac{\sqrt{a}}{\sqrt{b}}$$

IN OTHER WORDS, NO SURPRISES!

TOO BAD... I **LIKE** SURPRISES...

(ASSUMING $a \geq 0$ AND $b > 0$, THAT IS!)

THE REASON IS THE SAME AS FOR PRODUCTS. (THIS ISN'T SURPRISING, BECAUSE QUOTIENTS ARE PRODUCTS IN DISGUISE, REALLY...) SO, SQUARING THE QUOTIENT ON THE RIGHT GIVES

I'M GOING TO A SURPRISE PARTY FOR MYSELF...

$$\left(\frac{\sqrt{a}}{\sqrt{b}}\right)^2 = \frac{(\sqrt{a})^2}{(\sqrt{b})^2}$$

BY THE RULE FOR MULTIPLYING FRACTIONS

$$= \frac{a}{b}$$

SO THE QUOTIENT $\sqrt{a}/\sqrt{b}$ IS THE SQUARE ROOT OF a/b.

Example 8. $\sqrt{\dfrac{3}{4}} = \dfrac{\sqrt{3}}{\sqrt{4}} = \dfrac{\sqrt{3}}{2}$

Example 9. $\sqrt{\dfrac{1}{9}} = \dfrac{\sqrt{1}}{\sqrt{9}} = \dfrac{1}{3}$

Example 10. $\sqrt{\dfrac{1}{b}} = \dfrac{1}{\sqrt{b}}$

Example 11. $\sqrt{\dfrac{1}{a^2}} = \dfrac{1}{|a|}$

DON'T WE DESERVE **SOMETHING** FOR LEARNING ALL THIS?

Radicals out of DENOMINATORS!

HERE'S A USEFUL LITTLE EQUATION—AND IT MAY EVEN SURPRISE YOU.

SINCE YOU SAID YOU LIKE SUR-PRISES!

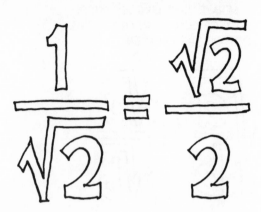

$$\frac{1}{\sqrt{2}} = \frac{\sqrt{2}}{2}$$

TO SEE THIS, SIMPLY MULTIPLY THE LEFT-HAND SIDE BY $\sqrt{2}/\sqrt{2}$. SINCE $\sqrt{2}/\sqrt{2} = 1$, THE MULTIPLICATION DOESN'T CHANGE THE EXPRESSION'S VALUE. IN THE END, THE RADICAL DISAPPEARS FROM THE DENOMINATOR.

$$\frac{1}{\sqrt{2}} = \frac{1}{\sqrt{2}} \frac{\sqrt{2}}{\sqrt{2}}$$

$$= \frac{\sqrt{2}}{2}$$

POOF! GONE!

THIS WORKS FOR ANY POSITIVE NUMBER OR EXPRESSION UNDER THE RADICAL SIGN, NOT JUST 2. IN OTHER WORDS, WE CAN **ALWAYS REMOVE A LONE RADICAL** FROM THE DENOMINATOR!!

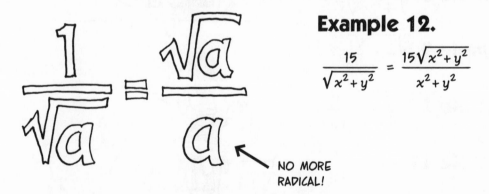

$$\frac{1}{\sqrt{a}} = \frac{\sqrt{a}}{a}$$

NO MORE RADICAL!

Example 12.

$$\frac{15}{\sqrt{x^2+y^2}} = \frac{15\sqrt{x^2+y^2}}{x^2+y^2}$$

PRODUCTS of SUMS

MAY BE SIMPLER THAN YOU THINK.

WHICH IS GOOD, BECAUSE I'M THINKING **VERY** COMPLICATED THOUGHTS...

Example 13. FIND $(3+\sqrt{2})(5+4\sqrt{2})$. TO DO THIS, WE MULTIPLY AS WE WOULD ANY PRODUCT OF SUMS.

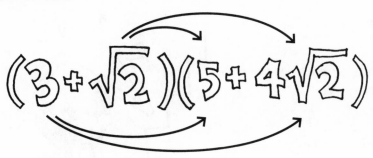

$(3+\sqrt{2})(5+4\sqrt{2})$

$= (3)(5)+5\sqrt{2}+(3)(4)\sqrt{2}+4\sqrt{2}\sqrt{2}$

$= 15+5\sqrt{2}+12\sqrt{2}+4(\sqrt{2})^2$

$= 15+17\sqrt{2}+(4)(2)$

$= \mathbf{23+17\sqrt{2}}$

SNEAKY!

ANOTHER SURPRISE!

THE ORIGINAL FOUR TERMS SHRANK DOWN TO TWO. THIS HAPPENED BECAUSE $\sqrt{2}$ WAS MULTIPLIED BY ITSELF, IN OTHER WORDS, **SQUARED**, MAKING 2... SO THE RADICAL SIGN BITES THE DUST...

MORE VANISHING RADICALS!

LOOK WHAT HAPPENS TO THE PRODUCT $(a + \sqrt{b})(a - \sqrt{b})$. IT'S $a^2 - (\sqrt{b})^2$, THAT IS

$$(a + \sqrt{b})(a - \sqrt{b}) = a^2 - b$$

Example 14a. $(5 + \sqrt{23})(5 - \sqrt{23}) = 25 - 23 = \mathbf{2}$

Example 14b. $(\sqrt{8} + \sqrt{7})(\sqrt{8} - \sqrt{7}) = 8 - 7 = \mathbf{1}$

THE BEAUTY OF THIS ONE IS THAT IT LETS US REMOVE RADICALS FROM DENOMINATORS EVEN WHEN THE RADICALS ARE COMBINED WITH OTHER TERMS, AS IN

$$\frac{1}{a + \sqrt{b}}$$

WE CLEAR THE RADICAL BY MULTIPLYING TOP AND BOTTOM BY $a - \sqrt{b}$.

$$\frac{1}{a + \sqrt{b}} = \frac{1}{a + \sqrt{b}} \cdot \frac{a - \sqrt{b}}{a - \sqrt{b}}$$

$$= \frac{a - \sqrt{b}}{a^2 - b}$$

Example 15. SIMPLIFY $\dfrac{1}{\sqrt{3} + \sqrt{2}}$.

SOLUTION: MULTIPLY NUMERATOR AND DENOMINATOR BY $\sqrt{3} - \sqrt{2}$.

$$\frac{1}{\sqrt{3} + \sqrt{2}} \cdot \frac{\sqrt{3} - \sqrt{2}}{\sqrt{3} - \sqrt{2}} = \frac{\sqrt{3} - \sqrt{2}}{(\sqrt{3})^2 - (\sqrt{2})^2}$$

$$= \frac{\sqrt{3} - \sqrt{2}}{3 - 2} = \mathbf{\sqrt{3} - \sqrt{2}}$$

THEY ALL BELONG UPSTAIRS!!

NOW THAT WE'VE PUT SQUARE ROOTS IN THEIR PLACE, LET'S REVIEW WHERE WE'VE BEEN...

IN THE PREVIOUS CHAPTER, WE HAD OUR FIRST LOOK AT QUADRATIC EXPRESSIONS AND THEIR ROOTS, VALUES OF x WHERE AN EXPRESSION IS ZERO... BUT FINDING THESE ROOTS REMAINED A MYSTERIOUS PROCESS.

SNUF SNUF

YOU NEED A NOSE FOR IT, I GUESS...

IN THIS CHAPTER, WE LOOKED AT THE SPECIAL ROOTS CALLED **SQUARE** ROOTS, AND WE LEARNED HOW TO ADD, MULTIPLY, AND DIVIDE THEM. SQUARE ROOTS ARE SPECIAL BECAUSE THEY SOLVE A SIMPLE EQUATION: $\sqrt{p}$ SOLVES THE EQUATION $x^2 = p$ OR $x^2 - p = 0$.

$\sqrt{p}$ AND $-\sqrt{p}$ ARE THE ROOTS OF $x^2 - p$!

IN THE NEXT CHAPTER, WE'LL SEE HOW TO FIND THE ROOTS OF **ANY** QUADRATIC EXPRESSION—IN TERMS OF SQUARE ROOTS. IN OTHER WORDS, WE WILL NEED TO USE THAT RADICAL SIGN! READ ON...

Problems

1. SIMPLIFY BY ADDING, SUBTRACTING, MULTIPLYING, DIVIDING, OR REMOVING SQUARES FROM UNDER THE RADICAL SIGN:

a. $\sqrt{64}$

b. $\sqrt{9+16}$

c. $3\sqrt{7} + 4\sqrt{7}$

d. $4 + \sqrt{3} - (2 - 3\sqrt{3})$

e. $(\sqrt{2})(2\sqrt{2})$

f. $\sqrt{\dfrac{1}{16}}$

g. $\dfrac{1}{\sqrt{2}} \cdot \dfrac{8}{\sqrt{2}}$

h. $\sqrt{5^3}$

i. $\sqrt{5^4}$

j. $(-\sqrt{2})(\sqrt{2})$

k. $(1 + \sqrt{5})(1 - \sqrt{5})$

l. $(\sqrt{3} + \sqrt{5})(1 + \sqrt{3})$

m. $\sqrt{\dfrac{4}{9}}$

n. $\sqrt{\dfrac{2}{9}}$

o. $\sqrt{(-4)(-4)}$

2. IF $\sqrt{3} \approx 1.73205081$ AND $3\sqrt{3} \approx 5.19615242$, THEN WHAT IS

$$\dfrac{5.19615242}{1.73205081}$$

APPROXIMATELY?

3. SHOW THAT $\sqrt{6} + \sqrt{24} = \sqrt{54}$.

4. SHOW THAT $\sqrt{8} + \sqrt{2} = 3\sqrt{2}$.

5. WHY IS $15 = \sqrt{45 \times 5}$?

6. SHOW HOW TO DRAW A LINE OF LENGTH $\sqrt{3}$ INCHES.

7. WHY IS $\sqrt{(m/n)} = \sqrt{|m|} / \sqrt{|n|}$?

8. WITHOUT DOING THE MULTIPLICATION, FIND $\sqrt{16 \times 25}$. WHAT IS 16×25 ?

9. SIMPLIFY $\sqrt{17} + \sqrt{68}$

10. IF $p = \dfrac{\sqrt{5} - 1}{2}$, SHOW THAT

$$p = \dfrac{1}{p+1}$$

11. WHAT ARE THE ROOTS OF $x^2 - 4$? OF $x^2 - 2$? OF $x^2 - 5$?

12. CLEAR RADICALS FROM DENOMINATORS:

a. $\dfrac{1}{\sqrt{3}}$

b. $\dfrac{5}{\sqrt{5}}$

c. $\dfrac{\sqrt{2}}{1 + \sqrt{2}}$

d. $\dfrac{2}{\sqrt{p+2} + \sqrt{p}}$

e. $\dfrac{1}{\sqrt{a} - \sqrt{b}}$

13a. EXPAND $(x + \sqrt{2})^2$.

13b. EXPAND $(x + \sqrt{a})^2$.

14. WHAT ARE THE ROOTS OF $(x - \sqrt{a})^2$?

15. IF a, b, d, c ARE INTEGERS, AND $n > 0$, SHOW THAT

$$(a + b\sqrt{n})(c + d\sqrt{n}) = p + q\sqrt{n},$$

WHERE p AND q ARE ALSO INTEGERS.

16. IF $0 < a < 1$, WHY IS $a^2 < a$? WHY IS $\sqrt{a} > a$?

17. USING A CALCULATOR, CHECK THAT

$$\dfrac{1}{\sqrt{3} + \sqrt{2}} = \sqrt{3} - \sqrt{2}.$$

WHAT IS THIS NUMBER, TO FIVE DECIMAL PLACES?

18. IF a, b, c, d ARE RATIONAL AND n IS A POSITIVE INTEGER, SHOW THAT

$$\dfrac{a + b\sqrt{n}}{c + d\sqrt{n}} = p + q\sqrt{n}$$

WHERE p AND q ARE BOTH RATIONAL.

Chapter 15
Solving Quadratic Equations

WE CAN SOLVE **ANY** QUADRATIC EQUATION, REALLY—
OR SOMETIMES, NOT SO REALLY...

AS WE'VE ALREADY MENTIONED, GIVEN AN EQUATION

$$ax^2 + bx + c = 0$$

IT'S OKAY TO DIVIDE BOTH SIDES BY a, SO WE'LL ASSUME FOR MOST OF THIS CHAPTER THAT THE COEFFICIENT OF x IS 1. WE'LL SOLVE THIS EQUATION FIRST:

$$x^2 + bx + c = 0$$

NO FAIR... I DON'T GET WRITTEN DOWN LIKE ALL THE OTHER COEFFICIENTS...

THAT'S BECAUSE YOU DON'T **DO** ANYTHING!!

Solving by **FACTORING**

ON P. 178, WE SAW THAT THE EQUATION

$$(x - r)(x - s) = 0$$

HAS TWO SOLUTIONS, r AND s, BECAUSE EACH OF THESE NUMBERS "ZEROES OUT" ONE OF THE FACTORS. THE SAME IS TRUE OF

$$(x + p)(x + q) = 0$$

EXCEPT NOW THE SOLUTIONS ARE −p AND −q, FOR THE SAME REASON.

YUP—LOOKS LIKE ZERO...

WE ALSO SAW THAT $(x + p)(x + q) = x^2 + (p+q)x + pq$. WHAT WE'RE HOPING NOW IS THAT, GIVEN AN EXPRESSION $x^2 + bx + c$, WE CAN "PULL IT APART" AND FIND **FACTORS** $x+p$ AND $x+q$ SO THAT $(x + p)(x + q) = x^2 + bx + c$. IF WE CAN, THEN IT MUST BE TRUE THAT

$$p + q = b \qquad pq = c$$

SOUNDS FAMILIAR...

FOR INSTANCE, GIVEN THE EXPRESSION $x^2 + 5x + 6$, IS THERE A PAIR OF NUMBERS THAT **ADD** TO **5** AND **MULTIPLY** TO **6**?

YOU MAY ALREADY SEE THAT THE NUMBERS 3 AND 2 DO THE TRICK.

$$3 + 2 = 5$$
$$3 \times 2 = 6$$

SO $(x+3)(x+2) = x^2 + 5x + 6$.

IN GENERAL, TO UNSCRAMBLE OR **FACTOR** A QUADRATIC EXPRESSION $x^2 + bx + c$, WE MUST FIND TWO NUMBERS WHOSE **PRODUCT** IS THE CONSTANT TERM c AND WHOSE **SUM** IS THE LINEAR COEFFICIENT b. THE BABYLONIAN PROBLEM LIVES!

YOU THINK **YOU'RE** IMMORTAL? I INVENTED BEER!

MORE EXAMPLES:

Example 1. FACTOR $x^2 + 4x + 3$.

STEP 1. THINK OF ALL WAYS TO FACTOR 3. LUCKILY, THERE'S ONLY ONE WAY:

$$3 = 3 \times 1$$

STEP 2. FIND THE SUM OF THE TWO FACTORS OF 3:

$$3 + 1 = 4$$

SINCE 4 IS THE COEFFICIENT OF x, THIS PAIR OF NUMBERS SOLVES THE PROBLEM.

$$x^2 + 4x + 3 = (x+1)(x+3)$$

AS YOU CAN EASILY CHECK BY EXPANDING THE RIGHT-HAND SIDE. THE ROOTS OF $x^2 + 4x + 3$ ARE -1 AND -3.

Example 2. FACTOR $x^2 + 11x + 24$.

STEP 1. THE CONSTANT TERM, 24, HAS SEVERAL FACTORIZATIONS:

$$24 = 1 \times 24$$
$$= 2 \times 12$$
$$= 3 \times 8 \quad \longleftarrow$$
$$= 4 \times 6$$

STEP 2. CHECK FOR A PAIR THAT SUMS TO 11, THE COEFFICIENT OF x. WE FIND

$$3 + 8 = 11$$

THIS SOLVES THE PROBLEM. THE EXPRESSION'S ROOTS ARE -3 AND -8, AND

$$x^2 + 11x + 24 = (x+3)(x+8)$$

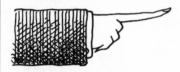 **First find factors of c, then check their sums.**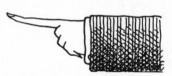

Example 3. FACTOR $x^2 - x - 6$.

HERE THE CONSTANT TERM IS NEGATIVE, SO IT MUST BE THE PRODUCT OF A **POSITIVE** FACTOR AND A **NEGATIVE** FACTOR.

STEP 1. LOOK AT FACTORS OF -6.

$$-6 = (1)(-6)$$
$$= (2)(-3) \longleftarrow$$
$$= (3)(-2)$$
$$= (6)(-1)$$

STEP 2. WE NEED A PAIR THAT SUMS TO THE COEFFICIENT OF x, WHICH IS **-1**. THE SECOND PAIR, 2, -3 DOES THE JOB: $2 - 3 = -1$, SO

$$x^2 - x - 6 = (x + 2)(x - 3)$$

Example 4. FACTOR $x^2 + 2x - 8$.

AGAIN THE CONSTANT TERM -8 IS NEGATIVE, SO WE HAVE TO CONSIDER ONE POSITIVE AND ONE NEGATIVE FACTOR.

STEP 1. LOOK AT FACTORS OF -8.

$$-8 = (1)(-8)$$
$$= (2)(-4)$$
$$= (4)(-2) \longleftarrow$$
$$= (8)(-1)$$

STEP 2. WE NEED A PAIR THAT SUM TO **2**. THE THIRD PAIR, 4, -2, WORKS: $4 - 2 = 2$, AND SO

$$x^2 + 2x - 8 = (x + 4)(x - 2)$$

Example 5. FACTOR $x^2 - 10x + 24$. HERE $c = 24 > 0$ BUT $b = -10 < 0$. THE FACTORS

OF 24 MUST BOTH BE POSITIVE OR BOTH BE NEGATIVE. BUT TWO POSITIVES CAN'T ADD TO -10, SO THE ONLY POSSIBILITY IS TWO NEGATIVE FACTORS.

1. WRITE 24 AS A PRODUCT OF NEGATIVE FACTORS.

$$24 = (-1)(-24)$$
$$(-2)(-12)$$
$$(-3)(-8)$$
$$(-4)(-6)$$

2. CHECKING THEIR SUMS, WE SEE THAT

$$-4 - 6 = -10$$

AND CONCLUDE THAT

$$x^2 - 10x + 24 = (x - 4)(x - 6)$$

I AM **SO** GLAD SOMEBODY THOUGHT OF NEGATIVE NUMBERS...

CLEARLY, IT'S IMPORTANT TO KEEP TRACK OF SIGNS WHEN FACTORING! WE CAN SPECIFY THE SIGNS OF p AND q WITH A "LOGIC TREE" SHOWING WHAT HAPPENS FOR EACH COMBINATION OF SIGNS OF b AND c.

SPECIFICALLY,

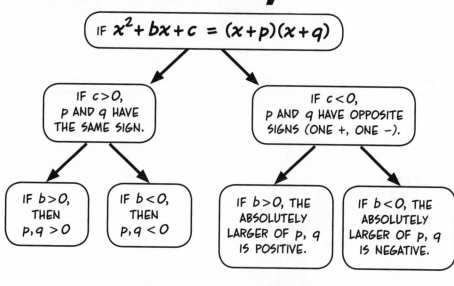

IF $x^2 + bx + c = (x+p)(x+q)$

IF $c > 0$, p AND q HAVE THE SAME SIGN.

IF $c < 0$, p AND q HAVE OPPOSITE SIGNS (ONE +, ONE −).

IF $b > 0$, THEN $p, q > 0$

IF $b < 0$, THEN $p, q < 0$

IF $b > 0$, THE ABSOLUTELY LARGER OF p, q IS POSITIVE.

IF $b < 0$, THE ABSOLUTELY LARGER OF p, q IS NEGATIVE.

TREE? IT LOOKS MORE LIKE AN ALIEN PROBE DEVICE...

USE YOUR IMAGINATION...

WE CAN ALSO SUMMARIZE THIS IN A TABLE. FOR SIMPLICITY, ASSUME THAT $|p| > |q|$. (I.E., p IS THE ABSOLUTELY LARGER OF THE TWO.)

c	b	
+	+	$p, q > 0$
+	−	$p, q < 0$
−	+	$p > 0,\ q < 0$
−	−	$p < 0,\ q > 0$

AND THEN THERE'S THIS!

Example 6. FACTOR $x^2 + 2x - 6$.

STEP 1. FROM THE LOGIC TREE, WE SEE THAT $p > 0$ AND $q < 0$. SO...

$$-6 = (-1)(6)$$
$$= (-2)(3)$$

STEP 2. DOES EITHER PAIR SUM TO 2, THE COEFFICIENT OF x?

$$6 - 1 = 5$$
$$3 - 2 = 1$$

UM... NO...

OOPS!

OUR STEP-BY-STEP METHOD HAS COME TO A DEAD STOP. **WHAT** ARE WE GOING TO DO?

THERE ARE AT LEAST TWO WAYS TO SOLVE THIS PROBLEM: THE BABYLONIAN WAY AND THE MODERN, ALGEBRAIC WAY. WE'LL SHOW THE ALGEBRAIC WAY AND LEAVE THE BABYLONIAN SOLUTION TO YOU AS A PROBLEM.

SEE YOU LATER!

A VERY SPECIAL EQUATION

LET'S TAKE WHAT MIGHT LOOK LIKE A SIDE
TRIP AND THINK ABOUT THIS EQUATION FOR
A FEW MINUTES.

$$(x + B)^2 = D$$

TRUE, WE HAVEN'T SEEN **EXACTLY** THIS EQUATION BEFORE, BUT
LET'S BLUNDER AHEAD AND TRY TO SOLVE IT ANYWAY...

MAYBE YOU
CAN SEE THAT
WE'RE GOING
TO NEED OUR
FRIEND, THE
SQUARE ROOT...

Example 7. SOLVE

$$(x - 3)^2 = 2$$

SOLUTION: SIMPLY TAKE THE SQUARE ROOT OF BOTH SIDES!

$$x - 3 = \pm\sqrt{2}$$

(COULD BE EITHER SQUARE ROOT.)

$$\boxed{x = 3 \pm \sqrt{2}}$$

ADDING 3 TO BOTH SIDES.

NOTE WELL! THIS IS REALLY **TWO** SOLUTIONS, IN ABBREVIATED FORM. IT MEANS THAT **BOTH** THESE VALUES SATISFY THE EQUATION:

 AND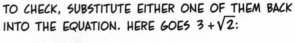

$$3 + \sqrt{2} \quad \text{AND} \quad 3 - \sqrt{2}$$

DON'T BELIEVE ME?

TO CHECK, SUBSTITUTE EITHER ONE OF THEM BACK INTO THE EQUATION. HERE GOES $3 + \sqrt{2}$:

$$((3 + \sqrt{2}) - 3)^2 \overset{?}{=} 2$$

$$(\sqrt{2})^2 \overset{?}{=} 2 \qquad \text{THE 3s CANCEL}$$

$$2 = 2$$

YOU SHOULD TRY SUBSTITUTING THE OTHER VALUE, $3 - \sqrt{2}$, TO SEE THAT IT GIVES THE SAME RESULT.

NOW LET'S DO THE SAME ALGEBRA ON THE GENERAL EQUATION $(x+B)^2 = D$.

$$(x+B)^2 = D$$
$$x+B = \pm\sqrt{D}$$
$$x = -B \pm \sqrt{D}$$

AND THERE YOU HAVE IT—OR **THEM**, REALLY!

AGAIN, TWO ANSWERS: $-B+\sqrt{D}$ AND $-B-\sqrt{D}$. YOU CAN CHECK THAT THEY BOTH SOLVE THE EQUATION BY PLUGGING THEM IN. THE $-B$ TERM CANCELS B, AND EITHER SQUARE ROOT ($+$ or $-$) SQUARES TO D.

WE'RE GETTING NEAR THE FINISH LINE!!

THERE'S STILL ONE LITTLE HURDLE... THIS WORKS **ONLY WHEN** D IS **NON-NEGATIVE**. OTHERWISE, WE'D BE TRYING TO TAKE THE SQUARE ROOT OF A NEGATIVE NUMBER, AND THAT IS A POSITIVE NO-NO, OR MAYBE A NEGATIVE ONE.

HM... WOULDN'T A NEGATIVE NO-NO BE "NO-NO-NO," WHICH EQUALS "NO"?

MAYBE...

Example 8. THE EQUATION

$$(x+5)^2 = -6$$

CAN'T BE SOLVED, AT LEAST BY ANY REAL NUMBER, BECAUSE $x+5$ WOULD HAVE TO BE $\sqrt{-6}$, AND WHAT'S **THAT?**

YOU MAY WONDER WHAT WE GET FROM SOLVING SUCH A SPECIAL EQUATION. HERE'S WHAT: IT TURNS OUT THAT WE CAN WRESTLE **EVERY QUADRATIC EQUATION** WITH LEADING COEFFICIENT 1 INTO THE FORM $(x+B)^2 = D$. THAT'S RIGHT! EVERY LAST ONE OF THEM. PERIOD. THIS BABYLONIAN TRICK IS CALLED...

Completing the SQUARE.

AGAIN, START WITH AN EXAMPLE. HERE'S THE THING WE COULDN'T FACTOR IN EXAMPLE 6.

Example 9. SOLVE $x^2 + 2x - 6 = 0$.

OUR PLAN IS TO TURN THIS INTO AN EQUATION LIKE $(x+B)^2 = D$. FIRST STEP: MOVE THE CONSTANT TERM TO THE RIGHT. NOW BOTH TERMS ON THE LEFT HAVE A FACTOR OF x.

$$x^2 + 2x = 6$$

BECAUSE $x^2 + 2x = x(x+2)$, WE CAN IMAGINE THE LEFT-HAND SIDE AS THE **AREA** OF A RECTANGLE WITH ONE SIDE x AND THE OTHER SIDE $x+2$.

WE'LL WORK WITH THE PART AT THE END THAT'S IN EXCESS OF x, AND MAKE THE BEST SQUARE WE CAN.

FIRST CUT OFF EXACTLY **HALF** THE STRIP. ITS WIDTH IS OBVIOUSLY 1, I.E., HALF OF 2.

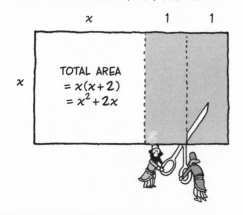

x 1   1

TOTAL AREA
$= x(x+2)$
$= x^2 + 2x$

MOVE THE SLICED-OFF BIT TO THE OTHER SIDE OF THE RECTANGLE.

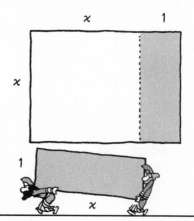

THE FIGURE BECOMES A LARGE SQUARE MISSING A SQUARE NOTCH OF SIDE 1.

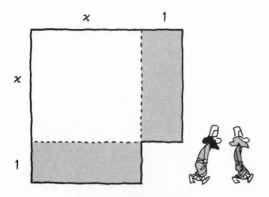

THE AREA REMAINS $x(x+2)$... UNTIL...

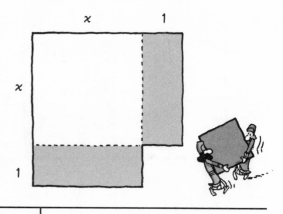

WE **COMPLETE THE SQUARE** BY FILLING IN THAT NOTCH. THIS ADDS AN AREA OF $1 \times 1 = 1$. THE TOTAL AREA—IT'S A SQUARE NOW!—IS $(x+1)^2$.

OR, ALGEBRA-ICALLY...

$$x(x+2) + 1 = (x+1)^2$$

ADDING 1 TO THE EQUATION'S LEFT SIDE MAKES IT A SQUARE. TO PRESERVE BALANCE, WE ADD 1 TO THE RIGHT ALSO.

$$x^2 + 2x + 1 = 6 + 1$$

$$(x+1)^2 = 7$$

AND THERE'S THE EQUATION IN THE FORM WE WANT! THE SOLUTIONS:

$$x = -1 \pm \sqrt{7}$$

CHECK BY SUBSTITUTING IN THE ORIGINAL EQUATION, OR, MORE EASILY, INTO $(x+1)^2 = 7$.

WE CAN COMPLETE THE SQUARE
OF **ANY** QUADRATIC EXPRESSION

JUST AS BEFORE, DRAW A RECTANGLE x BY
$x+b$. THE AREA IS $x(x+b) = x^2+bx$.

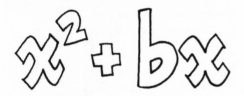

WITH NO CONSTANT TERM AND
LEADING COEFFICIENT 1. THE
STEPS ARE EXACTLY THE SAME.

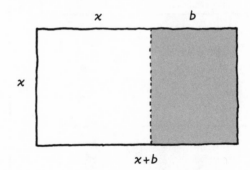

TEAR OFF A STRIP OF WIDTH $b/2$ AND MOVE IT AROUND TO MAKE A LARGE SQUARE MINUS A
SMALL SQUARE.

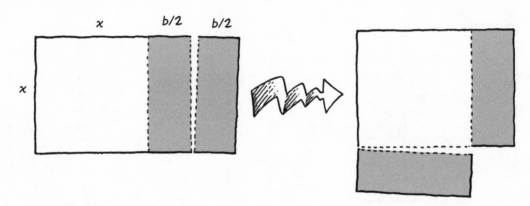

WE SEE THAT THE ORIGINAL AREA x^2+bx IS
"COMPLETED" TO $(x+b/2)^2$ BY ADDING A
SMALL SQUARE OF AREA $(b/2)^2 = b^2/4$.

THAT IS, COMPLETE THE SQUARE BY ADDING
THE **SQUARE OF HALF THE LINEAR CO-
EFFICIENT**, $(b/2)^2$ OR $\mathbf{b^2/4}$. IN SYMBOLS,

$$(x^2 + bx) + \frac{b^2}{4} = \left(x + \frac{b}{2}\right)^2$$

THE ADDED
TERM

THE PERFECT SQUARE

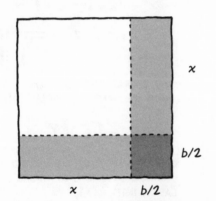

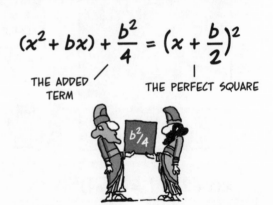

Example 10. COMPLETE $x^2 - 12x$.

SOLUTION: HALF OF 12 IS 6. ADDING 6^2 OR 36 GIVES

$$x^2 - 12x + 36 = (x - 6)^2$$

VISUALLY, WE'D GO THROUGH THE SAME STEPS WITH THIS RECTANGLE:

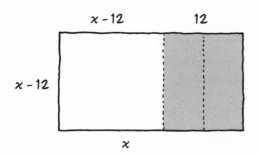

NOW LET'S LOSE THE RECTANGLE AND DO PURE ALGEBRA!

O-KAAYYYY...

Example 11.

TO **SOLVE** AN EQUATION (WITH LEADING COEFFICIENT 1) BY COMPLETING THE SQUARE:

$$x^2 - 6x + 4 = 0$$

1. MOVE THE CONSTANT TERM TO THE RIGHT.

$$x^2 - 6x = -4$$

2. COMPLETE THE SQUARE BY ADDING $b^2/4$ TO BOTH SIDES. HERE THAT'S $36/4 = 9$.

$$x^2 - 6x + 9 = 9 - 4$$

3. WRITE THE LEFT SIDE AS A SQUARE.

$$(x - 3)^2 = 9 - 4$$

4. SOLVE AS IN EXAMPLES 7 AND 9.

$$(x - 3)^2 = 5$$

$$x - 3 = \pm\sqrt{5}$$

$$x = 3 \pm \sqrt{5}$$

ALMOST THERE NOW!

RUNNING THROUGH THOSE SAME FOUR STEPS USING b AND c INSTEAD OF SPECIFIC NUMBERS AS IN THE LAST EXAMPLE, WE CAN SOLVE **ALL** QUADRATIC EQUATIONS (EXCEPT FOR THE ONES THAT WE CAN'T...) WITH THESE

QUADRATIC FORMULA(s).

GIVEN THIS EQUATION, WE FOLLOW THE RECIPE.

$$x^2 + bx + c = 0$$

STEP 1. MOVE THE CONSTANT...

$$x^2 + bx = -c$$

STEP 2. COMPLETE THE SQUARE BY ADDING $b^2/4$ TO BOTH SIDES...

$$x^2 + bx + \frac{b^2}{4} = \frac{b^2}{4} - c$$

$$= \frac{b^2 - 4c}{4}$$

STEP 3. EXPRESS THE LEFT-HAND SIDE AS A SQUARE.

$$\left(x + \frac{b}{2}\right)^2 = \frac{b^2 - 4c}{4}$$

STEP 4. SOLVE!!

$$x + \frac{b}{2} = \pm\sqrt{\frac{b^2 - 4c}{4}}$$

TAKING THE SQUARE ROOT

$$= \pm\frac{\sqrt{b^2 - 4c}}{2}$$

REMOVING THE RADICAL FROM THE DENOMINATOR

Conclusion: THE ROOTS ARE

(1) $$x = \frac{-b \pm \sqrt{b^2 - 4c}}{2}$$

SO LONG AS $b^2 - 4c \geq 0$, ANYWAY...

What if a≠1?

AND WHAT IF THE LEADING COEFFICIENT ISN'T 1? WHAT IF WE'RE FACING... THIS?

$$ax^2 + bx + c = 0$$

NOT A PROBLEM! DIVIDING THROUGH BY a, WE SEE THAT THIS HAS THE SAME SOLUTIONS AS

$$x^2 + (b/a)x + (c/a) = 0$$

NOW THE LEADING COEFFICIENT IS 1, SO WE CAN USE THE QUADRATIC EQUATION 1, REPLACING b BY b/a AND c BY c/a. IF YOU WORK OUT THE ALGEBRA—WHICH YOU SHOULD!—YOU WILL FIND THESE ROOTS:

$$(2) \quad x = \frac{-b \pm \sqrt{b^2 - 4ac}}{2a}$$

THIS IS **THE** QUADRATIC FORMULA MEMORIZED BY COUNTLESS GENERATIONS OF ALGEBRA STUDENTS... AND WHY SHOULD YOU BE ANY DIFFERENT?

WE'RE THERE! WE'RE THERE!

KUDOS! AND FOR YOUR PRIZE—A GIGANTIC FORMULA!

Example 12. SOLVE $2x^2 - 5x + 3 = 0$.

SOLUTION: COMPLETELY MINDLESSLY (THAT'S THE BEAUTY OF IT!) PLUG THE COEFFICIENTS INTO THE FORMULA. HERE $a = 2$, $b = -5$, $c = 3$. WE GET

$$\frac{5 \pm \sqrt{5^2 - (4)(3)(2)}}{(2)(2)} = \frac{5 \pm \sqrt{25 - 24}}{4}$$

$$= \frac{5}{4} \pm \frac{1}{4}$$

THAT IS, $\dfrac{3}{2}$ AND 1.

AHH... IT'S ALL OVER BUT THE HYDRATING!

UM... NOT QUITE...

WE SHOULD CHECK THE ANSWER, BY PLUGGING EACH ROOT BACK INTO THE QUADRATIC EXPRESSION AND EVALUATING— SHOULDN'T WE?

OOG... $(3/2)^2$ IS $9/4$... TIMES 2... OH, BROTHER...

IN FACT, **NO!!** WITH QUADRATICS, THERE'S ANOTHER, QUICKER WAY. TO CHECK THAT A PAIR r AND s ARE ROOTS OF THE QUADRATIC EXPRESSION $ax^2 + bx + c$, IT'S ENOUGH TO CHECK THAT

$$r + s = -\frac{b}{a} \text{ AND}$$
$$rs = \frac{c}{a}$$

A TOTALLY FABULOUS LABOR-SAVER!!

WELL, HEY!

208

WHY? WELL, IT'S CERTAINLY TRUE THAT r AND s ARE ROOTS OF $(x-r)(x-s)$... AND WE KNOW THAT

$$(x-r)(x-s) = x^2 - (r+s)x + rs$$

SO... IF r AND s SATISFY THE "BABYLONIAN EQUATIONS" $r+s = -b/a$ AND $rs = c/a$, THEN

$$(x-r)(x-s) = x^2 + \frac{b}{a}x + \frac{c}{a}$$

THIS SHOWS THAT r, s ARE ROOTS OF

$$x^2 + \frac{b}{a}x + \frac{c}{a}$$

SO THEY'RE ALSO ROOTS OF

$$ax^2 + bx + c.$$

CHECKING ROOTS IS NOW OFFICIALLY EASY!

THIS MEANS THAT THE ORIGINAL EXPRESSION HAS THIS FACTORIZATION:

$$ax^2 + bx + c = a(x-r)(x-s)$$

LET'S CHECK THE ANSWERS IN EXAMPLE 12 IN THIS WAY. THE ROOTS' SUM SHOULD BE $-(-5)/2 = 5/2$, AND THEIR PRODUCT SHOULD BE 3/2. AND IN FACT:

$$\frac{3}{2} + 1 = \frac{5}{2} \qquad \left(\frac{3}{2}\right)\cdot 1 = \frac{3}{2} \quad \text{CHECK!}$$

WE CONCLUDE THAT THE EXPRESSION CAN BE FACTORED AS

$$2\left(x - \frac{3}{2}\right)(x - 1) = (2x - 3)(x - 1)$$

ANYTHING ELSE?

The **DISCRIMINANT**

THE QUADRATIC FORMULA'S SQUARE-ROOT TERM $\sqrt{b^2 - 4ac}$ RAISES A KNOTTY PROBLEM: THE STUFF IN THERE MIGHT BE NEGATIVE!

> JUST MOPPING UP NOW...

THIS QUANTITY $b^2 - 4ac$ IS CALLED THE EXPRESSION'S **DISCRIMINANT.** ITS SIGN **DISCRIMINATES** BETWEEN EXPRESSIONS THAT HAVE REAL ROOTS AND THOSE THAT DON'T.

$$x^2 + 5x + 7$$

$b^2 - 4ac = -3 < 0$

$$3x^2 - 2x - 3$$

$b^2 - 4ac = 40 > 0$

WHEN $b^2 - 4ac > 0$, ALL IS GOOD. THE QUADRATIC FORMULA GIVES US TWO REAL ROOTS, AND WE BREATHE A SIGH OF RELIEF...

WHEN $b^2 - 4ac = 0$, THE "TWO" ROOTS ARE

$$-b/2a + 0 \quad \text{AND} \quad -b/2a - 0$$

IN OTHER WORDS, "BOTH" ROOTS ARE $-b/2a$, AND THE ORIGINAL EXPRESSION FACTORS AS

$$a\left(x + \frac{b}{2a}\right)^2$$

IN WHICH CASE WE SAY IT HAS A **Double root.**

Example 13, a Double Root.

FIND THE ROOTS OF $4x^2 - 12x + 9$.

SOLUTION: TO APPLY THE QUADRATIC FORMULA, WE FIRST CALCULATE THE DISCRIMINANT.

$$b^2 - 4ac = (-12)^2 - (4)(4)(9)$$

$$= 144 - 144 = 0$$

THE "TWO" ROOTS ARE BOTH $-b/2a = 12/8 = 3/2$, AND YOU CAN EASILY CHECK THAT THE ORIGINAL EXPRESSION IS

$$4\left(x - \frac{3}{2}\right)^2 = (2x - 3)^2$$

SO IT'S NOT A MIRAGE...

NOPE NOPE!

TO SUMMARIZE!

THE DISCRIMINANT GIVES US THIS INFORMATION:

$b^2 - 4ac > 0$ TWO REAL ROOTS

$b^2 - 4ac = 0$ DOUBLE ROOT, SQUARE EXPRESSION

$b^2 - 4ac < 0$ NO REAL ROOTS

AND WHAT DO WE DO IF THERE ARE NO REAL ROOTS? JUST GIVE UP AND QUIT?

STOP? **NO WAY!!**

UM... CAN WE TAKE A VOTE...?

211

Imaginary Square Roots?

WHAT IF WE DIDN'T STOP WHEN WE HIT A NEGATIVE DISCRIMINANT? WHAT IF WE PRETENDED IT WAS OKAY AND JUST KEPT SOLVING? THAT'S WHAT SOME ITALIAN MATHEMATICIANS DID BACK IN THE DAY, AND THE RESULTS WERE... WELL... PRETTY GOOD!

BACK IN **Example 8,** THE EQUATION $(x + 5)^2 = -6$ OR $x^2 + 10x + 31 = 0$ BROUGHT US FACE-TO-FACE WITH $\sqrt{-6}$. THEN WE STOPPED... BUT NOW LET'S CRANK AWAY JUST AS IF $\sqrt{-6}$ WERE ANY OTHER NUMBER. (NOTE THAT HERE $b = 10$ AND $c = 31$.)

THE "SOLUTIONS" ARE

$$r = -5 - \sqrt{-6}, \; s = -5 + \sqrt{-6}$$

AND WE READILY CHECK THAT

$$r + s = -10 = -b$$

$$rs = (-5)^2 - (\sqrt{-6})^2$$

$$= 25 - (-6))$$

$$= 31 = c$$

IN OTHER WORDS, THE ROOTS BEHAVE JUST LIKE REAL ROOTS. WE JUST DON'T KNOW WHAT THEY **MEAN!**

THIS LED THE MATHEMATICAL WORLD TO ADOPT A **NEW NUMBER,** $\sqrt{-1}$. THIS THING, WHICH IS WRITTEN i FOR **IMAGINARY,** HAS THE UNSETTLING PROPERTY THAT $i^2 = -1$. OTHER THAN THAT, i OBEYS ALL THE USUAL LAWS OF ADDITION AND MULTIPLICATION. SO, FOR INSTANCE,

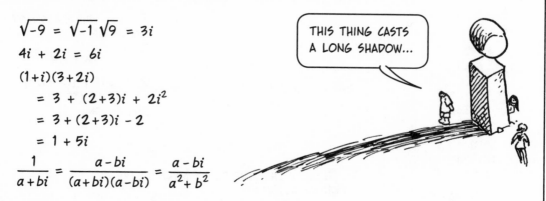

THIS THING CASTS A LONG SHADOW...

$$\sqrt{-9} = \sqrt{-1}\sqrt{9} = 3i$$

$$4i + 2i = 6i$$

$$(1+i)(3+2i)$$

$$= 3 + (2+3)i + 2i^2$$

$$= 3 + (2+3)i - 2$$

$$= 1 + 5i$$

$$\frac{1}{a+bi} = \frac{a-bi}{(a+bi)(a-bi)} = \frac{a-bi}{a^2+b^2}$$

THIS WORKS SO MAGICALLY WELL THAT i HAS BECOME A KEY PART OF ALL MODERN MATH. THE NUMBER i IS OFTEN THOUGHT OF AS A POINT ON A PLANE, NOT A LINE, AND MULTIPLICATION BY i AS A QUARTER-CIRCLE ROTATION AROUND THE ORIGIN.

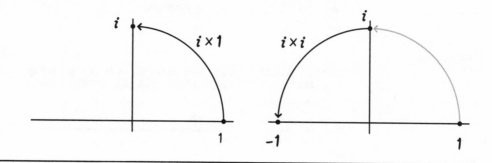

NUMBERS THAT COMBINE REALS AND "IMAGINARIES," LIKE $4 + 7i$ OR $2.7186 - 98.10107i$, ARE CALLED **COMPLEX** NUMBERS... AND BELIEVE IT OR NOT, IN SOME STRANGE, DEEP SENSE, THE REAL WORLD IS BEST DESCRIBED BY COMPLEX NUMBERS... AND THAT'S THE LAST THING I'LL HAVE TO SAY ABOUT THAT, IN THIS BOOK ANYWAY!!

'BYE, i!!

NOW IT'S TIME TO WORK ON SOME **REAL** PROBLEMS...

Problems

1. FACTOR.

a. $x^2 + 4x + 3$

b. $x^2 + 4x + 4$

c. $x^2 - 2x - 24$

d. $x^2 + 8x + 15$

e. $x^2 - 7x + 12$

f. $x^2 + 2x - 224$

g. $x^2 - x - 380$

2. SOLVE BY FACTORING. CHECK YOUR ANSWERS.

a. $x^2 - 4x + 3 = 0$

b. $x^2 + 15x + 26 = 0$

c. $x^2 + x - 6 = 0$

d. $x^2 - 4x - 5 = 0$

e. $x^2 + 9x + 20 = 0$

3. COMPLETE THE SQUARE OF EACH EXPRESSION.

a. $x^2 - 4x$

b. $x^2 - 6x$

c. $x^2 + x$

d. $x^2 + 9x$

e. $x^2 - 4\sqrt{5}x$

4. FIND THE DISCRIMINANT. IS THE EXPRESSION A PERFECT SQUARE? A CONSTANT MULTIPLE OF A PERFECT SQUARE? WHICH HAVE NO REAL ROOTS?

a. $x^2 + 4x + 3$

b. $2x^2 + 8x + 8$

c. $x^2 + x - 6$

d. $3x^2 - 4x + 5$

e. $x^2 + 9x + 20$

f. $x^2 + 10x + 25$

g. $x^2 + \frac{7}{2}x + 25$

5. SOLVE BY THE QUADRATIC EQUATION **AND** BY COMPLETING THE SQUARE. (TO COMPLETE THE SQUARE, DIVIDE BY THE LEADING COEFFICIENT IF NECESSARY.)

a. $3x^2 + 9x - 1 = 0$

b. $x^2 - 7x + 12 = 0$

c. $x^2 - x - 100 = 0$

d. $9x^2 + 10x + 1 = 0$

e. $x^2 - \sqrt{3}x - \frac{3}{2} = 0$

6. IF $i^2 = -1$, SHOW THAT

$$\frac{1 + i}{1 - i} = i$$

7. SHOW WHY 54 IS **NOT** A ROOT OF

$$x^2 - 73x + 1{,}027$$

WITHOUT PLUGGING IT IN TO EVALUATE THE EXPRESSION.

8. SHOW THAT THE ROOTS GIVEN BY THE QUADRATIC FORMULA,

$$r = \frac{-b + \sqrt{b^2 - 4ac}}{2a} \qquad s = \frac{-b - \sqrt{b^2 - 4ac}}{2a}$$

ADD TO $-b/a$ AND MULTIPLY TO c/a.

9. SINCE ANCIENT TIMES, TWO POSITIVE NUMBERS p AND q HAVE BEEN SAID TO HAVE THE **GOLDEN RATIO** IF

$$\frac{p}{q} = \frac{q}{p+q}$$

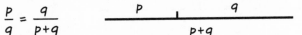

THAT IS, THE RATIO OF THE SMALLER TO THE LARGER IS THE SAME AS THE RATIO OF THE LARGER TO THE SUM. THE GREEKS BELIEVED THAT THE **GOLDEN RECTANGLE**, WITH SIDES IN THE GOLDEN RATIO, WAS THE MOST BEAUTIFUL OF ALL RECTANGLES.

a. SHOW THAT IF p, q HAVE THE GOLDEN RATIO (WITH $p < q$), THEN

$$\frac{q - p}{p} = \frac{p}{q}$$

IN OTHER WORDS, IF YOU REMOVE A SQUARE FROM A GOLDEN RECTANGLE'S END, THE REMAINING RECTANGLE IS GOLDEN!

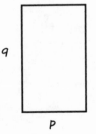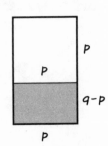

b. IF $p = 1$, FIND q. (HINT: MAKE A QUADRATIC EQUATION IN q.)

10. SOLVE THE "BABYLONIAN PROBLEM" DIRECTLY. THAT IS, GIVEN ANY TWO NUMBERS b AND c, FIND r AND s TO SATISFY THE EQUATIONS

$$r + s = b$$
$$rs = c$$

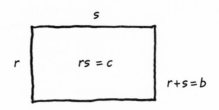

STEP 1. START WITH A RECTANGLE WITH SIDES r AND s AND AREA $rs = c$. LET $p = (s-r)/2$. PULL A STRIP OF LENGTH p FROM ONE SIDE AND PASTE IT ONTO THE OTHER TO MAKE A "NOTCHED SQUARE," STILL WITH AREA c. ITS LONG SIDE IS $r+p$ OR $s-p$.

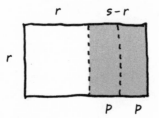

STEP 2. NOTE THAT THE MISSING PIECE IS A SQUARE OF SIDE p.

STEP 3. **(MOST IMPORTANT!)** SHOW THAT

$$r + p = s - p = \frac{r+s}{2} = \frac{b}{2}$$

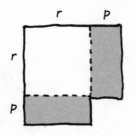

STEP 4. CONCLUDE THAT

$$\left(\frac{b}{2}\right)^2 = c + p^2$$

STEP 5. SOLVE FOR p IN TERMS OF b AND c.

STEP 6. FROM STEP 3, FIND

$$r = \frac{b}{2} + p \quad \text{AND}$$

$$s = \frac{b}{2} - p$$

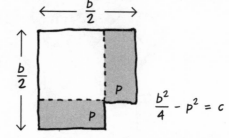

$$\frac{b^2}{4} - p^2 = c$$

STEP 7. FINALLY, EXPRESS r AND s IN TERMS OF b AND c. LOOK FAMLIAR?

11. SHOW THAT IF $x^2 + bx + c$ IS A SQUARE, THEN SO IS $cx^2 + bx + 1$.

12. FIND THE DISCRIMINANT OF THE EQUATION

$$(x+B)^2 = D$$

13. SOLVE THE BABYLONIAN PROBLEM PURELY ALGEBRAICALLY LIKE THIS: REPLACE r AND s WITH TWO NEW VARIABLES p AND q SO THAT

$$r = p + q \qquad s = p - q$$

THEN THE ORIGINAL EQUATIONS BECOME

$$2p = b \qquad p^2 - q^2 = c$$

FROM THESE FIND p AND q, AND FROM p AND q FIND r AND s.

Chapter 16
What's Next?

IN THIS BOOK, YOU'VE ACQUIRED THE BASIC TOOLS OF ALGEBRA.

STARTING WITH NUMBERS AND THEIR OPERATIONS, WE INTRODUCED THE IDEA OF A VARIABLE... AND THEN COMBINED VARIABLES AND NUMBERS INTO THE STUFF OF OUR STUDY, ALGEBRAIC EXPRESSIONS.

USING THE RULES OF ARITHMETIC, WE LEARNED HOW TO PUSH EXPRESSIONS AROUND WITHOUT CHANGING THEIR VALUE.

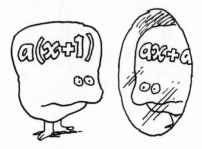

THIS LED US TO THE SOLUTION OF ALGEBRAIC EQUATIONS BY REBALANCING, COMBINING TERMS, AND SO ON.

WE DREW PICTURES OF EQUATIONS AND SOLVED PAIRS OF EQUATIONS IN TWO VARIABLES.

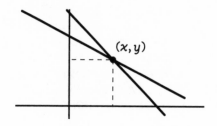

NEXT, WE USED VARIABLES IN DENOMINATORS TO STUDY PROPORTIONS, RATES, AND AVERAGES.

$$r = \frac{U - U_0}{t - t_0}$$

FINALLY, WE EXPLORED THE MYSTERIES OF SQUARES, SQUARE ROOTS, AND QUADRATIC EQUATIONS.

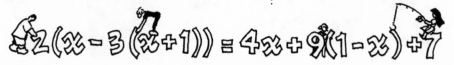

$$2(x - 3(x+1)) = 4x + 9(1 - x) + 7$$

SO...

WHAT ELSE IS THERE?

FIRST OF ALL, THERE ARE ALL THE USES OF ALGEBRA IN THE WORLD, FROM COMPUTER GRAPHICS TO HANDLING MONEY TO DESIGN, BUILDING, ENGINEERING, SIGNAL PROCESSING (IN TV, RADIO, MUSIC), AND MANY OTHER APPLICATIONS.

THEN THERE ARE ALL THE AREAS OF MATH YET TO COME. TO PURSUE ALMOST ANY OF THESE, YOU NEED TO BE COMFORTABLE WITH ALGEBRA.

HIDDEN TRUTHS

CATEGORY THEORY

ALGEBRAIC TOPOLOGY

DIFFERENTIAL GEOMETRY

NUMBER THEORY

CALCULUS

TOPOLOGY

FUNCTIONS

TRIGONOMETRY

SOLID GEOMETRY

PLANE GEOMETRY

AND THERE'S LOTS MORE ALGEBRA, TOO!

FOR ONE THING, WE CAN DRAW QUADRATIC EQUATIONS, JUST AS WE DREW LINEAR ONES.

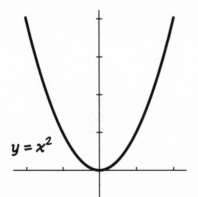

$y = x^2$

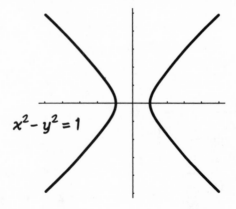

$x^2 - y^2 = 1$

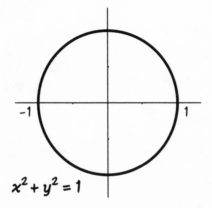

$x^2 + y^2 = 1$

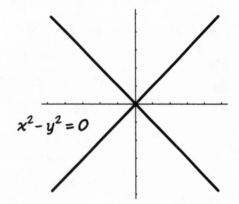

$x^2 - y^2 = 0$

ALGEBRA ALSO STUDIES **POLYNOMIALS** OF ANY DEGREE. (A POLYNOMIAL IS A SUM OF MANY TERMS OF DIFFERENT DEGREES.) THERE'S A LOT TO BE LEARNED FROM POLYNOMIALS AND THEIR GRAPHS!

I ASK YOU, WHAT'S MORE INTERESTING, A LINE OR A CURVE?

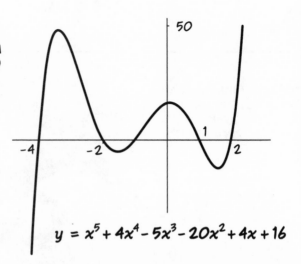

$y = x^5 + 4x^4 - 5x^3 - 20x^2 + 4x + 16$

EVEN **BINOMIALS**—TWO-TERM EXPRESSIONS LIKE $a + b$—ARE WORTH STUDYING. WHEN YOU RAISE THEM TO A POWER, LIKE $(a+b)^n$, THE COEFFICIENTS MAKE **PASCAL'S** BEAUTIFUL **TRIANGLE,** IN WHICH EACH NUMBER IS THE SUM OF THE TWO JUST ABOVE IT.

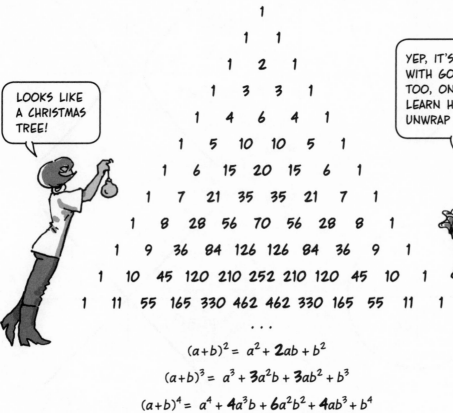

LOOKS LIKE A CHRISTMAS TREE!

YEP, IT'S LOADED WITH GOODIES, TOO, ONCE YOU LEARN HOW TO UNWRAP 'EM!

```
                    1
                 1     1
              1     2     1
           1     3     3     1
        1     4     6     4     1
     1     5    10    10     5     1
   1    6    15    20    15     6     1
  1    7    21    35    35    21    7    1
 1   8   28   56   70   56   28   8   1
1   9   36   84  126  126   84   36   9   1
1  10  45  120 210  252  210  120  45   10   1
1  11  55  165 330  462  462  330 165  55  11   1
                   . . .
```

$$(a+b)^2 = a^2 + 2ab + b^2$$

$$(a+b)^3 = a^3 + 3a^2b + 3ab^2 + b^3$$

$$(a+b)^4 = a^4 + 4a^3b + 6a^2b^2 + 4ab^3 + b^4$$

$$(a+b)^5 = a^5 + 5a^4b + 10a^3b^2 + 10a^2b^3 + 5ab^4 + b^5$$

ETC.

PASCAL'S TRIANGLE PLAYS A KEY ROLE IN MANY AREAS, INCLUDING THE LAWS OF **PROBABILITY.**

PROBABILITY? I INVENTED THAT!

PASCAL, OF COURSE!

ALGEBRA ALSO STUDIES
SEQUENCES, STRINGS
OF NUMBERS GENERATED BY SOME RULE. **ARITHMETIC** (ACCENT ON "MET") SEQUENCES ARE MADE BY ADDING THE SAME NUMBER AGAIN AND AGAIN.

$$a \quad a+b \quad a+2b \quad a+3b \quad a+4b \ldots$$

GEOMETRIC SEQUENCES COME FROM REPEATED MULTIPLICATION.

$$a, \; ar, \; ar^2, \; ar^3, \; \ldots$$

FOR INSTANCE, WHEN $a=1$ AND $r=\frac{1}{2}$,

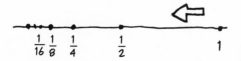

$$\tfrac{1}{16} \quad \tfrac{1}{8} \quad \tfrac{1}{4} \quad \tfrac{1}{2} \quad 1$$

SERIES ARE SUMS OF SEQUENCES.
ALGEBRA DISHES UP NICE FORMULAS FOR THESE.

$$1+2+3+\ldots+n = \frac{n(n+1)}{2}$$

$$1+r+r^2+\ldots+r^n = \frac{r^{n+1}-1}{r-1}$$

THE SECOND EQUATION, BY THE WAY, SHOWS THAT ADDING POWERS OF 2 GIVES THE NEXT POWER OF 2, LESS 1.

$$1+2+2^2+\ldots+2^n = 2^{n+1}-1$$

LINEAR ALGEBRA PLAYS WITH EQUATIONS IN MANY VARIABLES, WHERE NO VARIABLE
HAS A POWER HIGHER THAN 1. THIS IS THE MATH OF **FLAT THINGS** IN **HIGHER-DIMENSIONAL SPACES.** ALL COMPUTER GRAPHICS IS BASED ON LINEAR ALGEBRA.

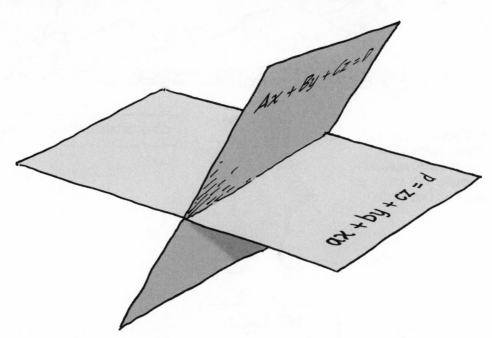

THEN THERE ARE THE POWERFUL BUT HIGHLY ABSTRACT SUBJECTS OF HIGHER ALGEBRA, LIKE **GROUP THEORY** AND **FIELD THEORY.** YOU GET THE PICTURE... THERE'S A LOT.

I'M NOT SURE I'M READY FOR THIS...

EVEN AT THE HIGHEST LEVEL, THOUGH, IT ALL RESTS ON A FOUNDATION OF BASIC ALGEBRA— IN OTHER WORDS, THE STUFF YOU JUST LEARNED IN THIS BOOK!

CONGRATULATIONS!

~THE END~

SOLUTIONS TO SELECTED PROBLEMS

Chapter 1, p. 12

1b. 93. **1c.** 1.5632. **1f.** 0.342 **1g.** 1.99996164 (ALMOST 2, IN OTHER WORDS!)

1i. 250 **2c.** 3.91666666... **2d.** 0.375 **2f.** 0.363636...

2g. 0.1764 7058 8235 2941 1764 7058 8235 2941 1764 7058 8235 2941 ... **2i.** 0.45

3. $3.91\overline{6}$ $0.\overline{36}$ $0.\overline{1764\ 7058\ 8235\ 2941}$ **4a.** $1\frac{1}{5}$ **4b.** $3\frac{2}{15}$ **5.** $\frac{3,514}{1,000}$

6.

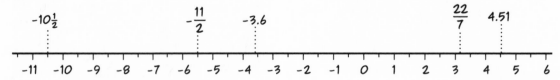

7b. 2 **7c.** -2 **7f.** $\frac{1}{2}$ **7h.** -22/7

8. THE VALUE IS 2 IF THE NUMBER OF MINUS SIGNS IS EVEN, AND -2 IF THE NUMBER OF MINUS SIGNS IS ODD.

Chapter 2, p. 22

1a. -27 **1d.** -1.1 **1f.** $-\frac{1}{6}$ **2b.** 19 **2d.** -12 **2f.** -2 **2g.** $-\frac{1}{48}$ **2i.** 98

4b. NEGATIVE **4c.** NEGATIVE **6.** HE "HAS" $(-13) **7b.** $-5-(-3)=-2$ **7c.** $16

Chapter 3, p. 34

1a. -27 **1c.** -24 **1f.** $\frac{1}{4}$ **1h.** 2 **1i.** 0 **2b.** 5 **2c.** 0

3. THE RECIPROCAL OF $-\frac{1}{3}$ IS -3. 0 HAS NO RECIPROCAL.   **4.** 50

6b. $\frac{1}{3}$ SITS ABOVE 1.

7. 3/2 LIES ABOVE 1.

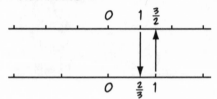

8.

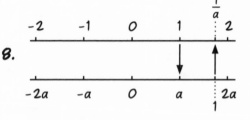

9.

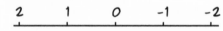

Chapter 4, p. 58

1a. 7 **1b.** 8 **1d.** 0 **1e.** 4 **1f.** $-\frac{1}{2}$ **1h.** $\frac{1}{3}$ **1j.** 50 **2b.** -1 **2d.** 0

3a. 9 **3c.** $10a - 10$ OR $10(a - 1)$ **4a.** $2x + 9$ **4d.** $13x + 9$ **4f.** $5a - 3at$

5. THE SALE PRICE IS $0.85P$.   **6.** THE THIRD AND FOURTH ROWS, FOR EXAMPLE, ARE

7. "RADDITION" IS ASSOCIATIVE AND COMMUTATIVE, BUT MULTIPLICATION DOES NOT DISTRIBUTE OVER "RADDITION."

$$(3 \times 2) \times 4 = 3 \times (2 \times 4)$$
$$(4 \times 2) \times 5 = 4 \times (2 \times 5)$$

8. ROTATION IS NOT COMMUTATIVE. IF P IS A POINT ON THE EQUATOR, AND R AND S ARE THE TWO ROTATIONS SHOWN, THEN DOING THEM IN ONE ORDER SENDS P TO THE NORTH POLE, WHILE ROTATING IN THE OPPOSITE ORDER PUTS P SOMEWHERE ELSE ON THE EQUATOR! HERE THE ORDER MATTERS.

IN THIS ORDER, P FIRST GOES AROUND THE EQUATOR, THEN TO THE NORTH POLE.

IN THE OPPOSITE ORDER, P NEVER LEAVES THE EQUATOR.

Chapter 5, p. 70

1b. $x = 3$ **1d.** $y = 5$ **1g.** $x = -\frac{1}{4}$ **1i.** $x = \frac{1}{3}$ **1l.** $t = \frac{5}{2}$ **1n.** $y = \frac{7}{4}$

2b. $\frac{3}{4}P$ **2c.** 88 **3a.** $p + .08p$ OR $(1.08)p$ **3c.** $(1 + r)p$ **4.** $x = 1/a$

5. EVERY NUMBER SOLVES THIS EQUATION, THANKS TO THE COMMUTATIVE LAW.

Chapter 6, p. 82

2. THE EQUATION IS $8(x + 2) = 10x$

3. THE EQUATION IS $8(x+3) - \dfrac{8(x+3)}{10} = 8x + \dfrac{8(x+3)}{10}$

KEVIN MAKES \$12/HR; JESSE MAKES \$15/HR.

5. THE EQUATION IS $2x + \dfrac{4x}{4} + 9 = 303$, AND THE FRAME IS 63" × 84".

7a. $5n$ **7c.** 7 NICKELS AND 14 DIMES **10.** \$590.40

Chapter 7, p. 94

1. $x = 27$, $y = 24$ **3.** $x = 1$, $y = 4$ **5.** $x = -27$, $y = 4$ **9.** $t = 3$, $u = -1$, $v = -2$

11a. $x = 14$, $y = 9$ **12.** 2,000 POUNDS OF BASS AND 3,000 POUNDS OF COD.

14. CELIA IS 14 AND JESSE IS 15.   **17.** $x = \dfrac{1}{2-a}$

Chapter 8, p. 114

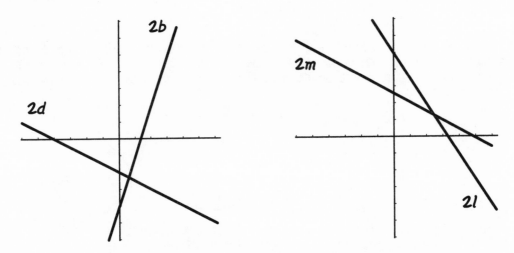

3a. $y = 3x + 5$ **3d.** $y = -\frac{1}{3}x - \frac{1}{5}$ **3e.** $y = -6x + 15$ **3g.** $y = 3x + 13$

4a. (3,4) IS ON THE LINE. (-3,1) IS NOT.   **4c.** (7,-2) LIES ON THE LINE.

WHEN $x = -14$, $y = 19$, SO THE TWO LINES MEET AT THE POINT (-14, 19).

5a. THE GIVEN GRAPH HAS SLOPE 4, SO THE EQUATION IS $y - 2 = 4(x - 1)$ OR $y = 4x + 6$.

5c. $y - 6.147 = -x + 2.35$ OR $x + y = 8.497$ **8.** $y_2 = y_1 + mp$

Chapter 9, p. 122

1c. $2^3 = 8$ **1d.** $2^{-4} = (1/16)$ **1g.** $(-2)^6 = 64$ **1i.** 3,125   **1l.** -196   **1m.** 21

1q. $\dfrac{1}{1,000,000}$ **1t.** 3   **1v.** 13   **2.** $(-6)^{100}$ IS POSITIVE. -6^{100} IS NEGATIVE.

4a. p^7 **4c** $6x^{50}$ **4g.** $-a^6x^3$ **4j.** a^{-n} OR $1/a^n$ **4k.** $32x^2$ **6.** 25 ZEROES

7d. 1.05×10^{13} **9.** 4,096

Chapter 10, p. 134

1a. 12   **1c.** 21   **1d.** 216   **1f.** 147   **2a.** $p^2 q^8$ **2c.** $4a^2 x^2 (x+1)$

2f. $(x-2)^2 (x+2)^3 (x+3)$ **2h.** $180(x^2+1)^3 (x^3-5)^4$

3b. $\dfrac{abx^2}{c^2}$

3c. $\dfrac{x^2+b^2}{bx}$

3e. $\dfrac{at^2 b^2}{3}$

4. $r = \dfrac{s}{sQ-1}$

5a. $\dfrac{a^2+t^2}{b^2}$

5c. $\dfrac{2(x+3)^2+(x+2)^2-6(x+1)^2}{(x+1)(x+2)(x+3)}$

5g. $\dfrac{B^2}{C}$

6c. 1,617

7. THEIR LCM MUST BE THEIR PRODUCT. THE REASON IS THAT THEY CAN SHARE NO COMMON FACTOR OTHER THAN 1. LET'S SEE WHY NOT.

IMAGINE, FOR INSTANCE, THAT 2 DIVIDES BOTH NUMBERS. THEN THEY'D BOTH BE EVEN, AND THEY MUST DIFFER BY AT LEAST 2.

IN GENERAL, CALL THE NUMBERS A AND B, AND ASSUME THAT THEY HAVE SOME COMMON FACTOR $p>1$. THEN $A=mp$ AND $B=np$ FOR SOME INTEGERS m AND n. THEIR DIFFERENCE, THEN, IS

$$A - B = mp - np$$
$$= p(m-n) \longleftarrow \text{ ITSELF A MULTIPLE OF } p,$$
$$\text{AND SO GREATER THAN 1.}$$

Chapter 11, p. 154

2. 3 GALLONS   **3.** 7/3 OUNCES PER MINUTE OR 1/6 PIECE PER MINUTE.   **5.** 23 OZ.

6b. IF L IS THE PORTION OF LAWN MOWED IN TIME t, THE EQUATION IS

$$L = \tfrac{1}{3}t + \tfrac{1}{2}\left(t - \tfrac{1}{2}\right)$$ AND THE WHOLE LAWN ($L=1$) IS MOWED IN AN HOUR AND A HALF.

7. $t = \dfrac{p+q}{pq}$

9. SET THE PROBLEM UP LIKE THIS: IMAGINE THE TWO POINTS A AND B ARE ON THE NUMBER LINE. WE CAN THEN LET ONE OF THEM BE THE ZERO POINT, THAT IS $A=0$. THEN THE TWO RUNNERS' VELOCITIES ARE

$$v_J = \dfrac{B \text{ ft}}{30 \text{ sec}} \qquad v_C = \dfrac{-B \text{ ft}}{25 \text{ sec}}$$

LETTING s BE POSITION, AS USUAL, THE RUNNERS' RATE EQUATIONS ARE

$$s_J = \dfrac{B(t - t_J)}{30} \qquad s_C = B - \dfrac{B(t - t_C)}{25}$$

WHERE t_J IS JESSE'S STARTING TIME, AND t_C IS CELIA'S STARTING TIME. WHEN THEY MEET, THESE POSITIONS ARE EQUAL.

PROBLEM 9, CONTINUED ⟋

IF THEY START AT THE SAME TIME, WE TAKE THAT TIME TO BE ZERO, SO THE EQUATION BECOMES $\frac{Bt}{30} = B - \frac{Bt}{25}$. B CANCELS, AND THE SOLUTION IS $t = 150/11$ SECONDS. IF CELIA STARTS 5 SECONDS AFTER JESSE, $t_c = 5$, AND THE EQUATION IS $\frac{Bt}{30} = B - \frac{B(t-5)}{25}$ AND THE SOLUTION IS $t = 180/11$ SEC.

13. PROBABLY NOT.

Chapter 12, p. 168

1a. 12   **1c.** 1,000,001   **1e.** $-\frac{3}{2}$ **1g.** 1   **1i.** 16   **2a.** 8   **2c.** 1   **2e.** A
2g. 793   **3.** MULTIPLYING BY $(a+b)(c+d)$ GIVES $a(c+d) = c(a+b)$, AND, AFTER EXPANDING, THE RESULT FOLLOWS.   **5.** 4 INCHES FROM THE SUSPENSION POINT ON THE SIDE OF THE SMALLER WEIGHT.   **7.** 48 MI/HR   **10.** YES, IT IS POSSIBLE! FOR INSTANCE:

	FIRST HALF	SECOND HALF	OVERALL
MOMO	3 FOR 4 = **.750**	30 FOR 100 = **.300**	33 FOR 104 = **.317**
JESSE	50 FOR 100 = **.500**	29 FOR 100 = **.290**	79 FOR 200 = **.395**

Chapter 13, p. 180

1.

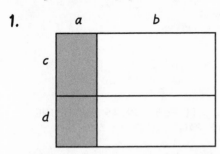

HERE $a(c+d)$ IS SHADED.

2a. $ab + 3a + 2b + 6$ **2c.** $6x^2 - 9x$
2e. $x^2 - 14x + 49$ **2g.** $6 - 5x + x^2$
3. $13 \times 17 = (15+2)(15-2) = 225 - 4 = 221$
4b. $1,000^2 - 5^2 = 1,000,075$
4c. $30^2 - 5^2 = 975$ **4e.** $1 - .0025 = .9975$
5b. 2 AND -5 **5d.** $-r$ AND $-s$ **5g.** 1, -3, 5
6b. 0   **8.** -17.458
9a. $4p^2 + qp^2 + 4q + q^2$
9b. $\frac{x^2}{2} + \frac{7x}{6} + \frac{2}{3}$ **9e.** $x^2 - x + \frac{1}{4}$
9i. $a^2x^2 + 2arx + r^2$ **9l.** $x^3 - 1$ **9n.** $x^5 + 1$

Chapter 14, p. 192

1b. 5   **1d.** $2 - 2\sqrt{3}$ **1f.** $\frac{1}{4}$ **1h.** $5\sqrt{5}$ **1j.** -2 **1l.** $3 + \sqrt{5} + \sqrt{3} + \sqrt{15}$ **1n.** $\frac{1}{3}\sqrt{3}$

2. 3   **3.** $\sqrt{24} = 2\sqrt{6}$, AND $3\sqrt{6} = \sqrt{3^2 \cdot 6} = \sqrt{54}$.   **4.** $\sqrt{(45)(5)} = \sqrt{3^2 \cdot 5^2} = 15$

6.

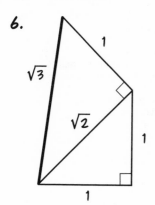

8. $\sqrt{16 \times 25} = 4 \times 5 = 20$, SO $16 \times 25 = 20^2 = 400$

12b. $\sqrt{5}$ **12c.** $-2 - \sqrt{2}$ **12e.** $\dfrac{\sqrt{a} + \sqrt{b}}{a + b}$

13b. $x^2 + 2\sqrt{a} + a$ **14.** ONLY ONE ROOT, $\sqrt{a}$

16. $a^2 < a$ BECAUSE a^2 IS a TIMES A POSITIVE NUMBER LESS THAN 1, NAMELY a ITSELF. $a < \sqrt{a}$ IS JUST ANOTHER WAY OF SAYING THE SAME THING.

18. MULTIPLY NUMERATOR AND DENOMINATOR BY $c - d\sqrt{n}$ AND COLLECT TERMS. THE RESULT CAN BE EXPRESSED AS

$$\frac{ac - bdn}{c^2 - nd^2} + \frac{bc - ad}{c^2 - nd^2}\sqrt{n}$$

BOTH THE FIRST TERM AND THE COEFFICIENT OF $\sqrt{n}$ ARE RATIONAL, BECAUSE SUMS, PRODUCTS, AND QUOTIENTS OF RATIONAL NUMBERS ARE RATIONAL.

Chapter 15, p. 214

1a. $(x + 3)(x + 1)$ **1c.** $(x - 6)(x + 4)$ **1f.** $(x + 16)(x - 14)$ **2b.** $x = -2$ AND $x = -13$

2d. $x = 5$ AND $x = -1$ **3b.** $x^2 - 6x + 9$ **3d.** $x^2 + 9x + \dfrac{81}{4}$ **3e.** $x^2 - 4\sqrt{5} + 20$

4b. 0. THE EXPRESSION IS TWICE THE SQUARE $x^2 + 4x + 4$.   **4d.** -44. NO REAL ROOTS

4g. $-87\frac{3}{4}$. NO REAL ROOTS   **5b.** ROOTS ARE 4 AND 3.   **5c.** $\frac{1}{2} \pm \frac{1}{2}\sqrt{401}$

7. IT CAN'T BE A ROOT, BECAUSE IF IT WERE, $73 - 54 = 19$ WOULD BE A ROOT ALSO, BUT $19 \times 54 \neq 1.027$.   **9a.** START BY BREAKING OUT THE FRACTION $(q-p)/p$ TO GET

$$\frac{q - p}{p} = \frac{q}{p} - 1.$$ THEN SUBSTITUTE $(p+q)/p$ FOR q/p (THE ORIGINAL ASSUMPTION) AND WORK OUT THE ALGEBRA.

11. THEY HAVE THE SAME DISCRIMINANT.   **12.** 4D

13. START WITH THE EQUATIONS $r = p + q$ FROM $b = r + s = 2p$ AND WORK
$s = p - q$ WHICH $c = rs = p^2 - q^2$ FROM THERE!

INDEX

About the Author

LARRY GONICK IS THE AUTHOR/CARTOONIST BEHIND THE AWARD-WINNING *CARTOON HISTORY OF THE UNIVERSE, THE CARTOON HISTORY OF THE UNITED STATES,* AND A STACK OF CARTOON GUIDES TO SCIENCE AND MATH. HE HAS TRAVELED THE WORLD IN SEARCH OF MATERIAL, AND, HAVING SEEN PLENTY, NOW MOSTLY STAYS HOME AND DRAWS. BEFORE TAKING UP PEN, BRUSH AND INKPOT, HE USED TO TEACH MATH AT HARVARD. HE'S MARRIED WITH CHILDREN.